Control of Infiltration and Inflow in Private Building Sewer Connections

Prepared by the **Sanitary Sewer Overflow Cooperative Agreement Workgroup of the Water Environment Federation**
Wayne Dillard, *Chair*

Under a cooperative agreement with the
U.S. Environmental Protection Agency
Assistance I.D. No. CX824505

1999

Water Environment Federation
601 Wythe Street
Alexandria, VA 22314-1994 USA

ISBN 1-57278-160-2
Printed in the USA **1999**

Water Environment Federation

Founded in 1928, the Water Environment Federation is a not-for-profit technical and educational organization. Its goal is to preserve and enhance the global water environment. Federation members number more than 42,000 water quality professionals and specialists from around the world, including engineers, scientists, government officials, utility and industrial managers and operators, academics, educators and students, equipment manufacturers and distributors, and other environmental specialists.

For information on membership, publications, and conferences, contact

Water Environment Federation
601 Wythe Street
Alexandria, VA 22314-1994
USA
1-703-684-2400

IMPORTANT NOTICE

PREFACE

This monograph will focus on private building sewer connections (PBSCs). Information will be presented that can be used by system operators to better understand the effects of infiltration/inflow (I/I) coming from PBSCs. An overview of operation, maintenance, and inspection practices will be included in the monograph that may help in the operation of the sewer system.

Finally, the monograph will present the results of a systemwide survey of PSC I/I, which will highlight the magnitude of I/I problems in private connections and methods and case histories addressing these problems.

Under a cooperative agreement with the U.S. Environmental Protection Agency, the Water Environment Federation conducted a national survey to generate information about sanitary sewer overflows. The goal of the survey is to raise awareness and provide educational, technical, and managerial information for the public and municipal officials; technical professionals; and owners, operators, and personnel of wastewater treatment plants.

Principal authors for this monograph are

Thomas J. Bagley
Dana L. Carroll
Andrew J. Davis
James C. Elliott
Steven Gee
Henry N. Gregory
Laurie Chase Mehl
Donna E. Renner
William H. Sukenik
William R. Weaver
Robert West
Dean Westman
William Zieburtz, Jr.

Review was provided by

Susan M. Dakak
David A. Hofer
Wen-Chi (Jim) Ku
Shellie R. Maxwell
James D. McGregor
Brendan O'Regan
Richard Prewitt
Allan L. Rae
Charles Vanderlyn
Chris Wilson

Authors' and reviewers' efforts were supported by the following organizations:

Black & Veatch, Atlanta, Georgia
Burns & McDonnell, Kansas City, Missouri
Camp Dresser & McKee, Woodbury, New York
Carroll Engineering & Associates, P.C., Belvidere, Illinois
City of Akron, Akron, Ohio
City of Columbus, Sewage & Drainage Division, Columbus, Ohio
City of Houston, Department of Public Works and Engineering, Houston, Texas
City of Newburyport, Sewer Department, Newburyport, Massachusetts
Dallas Area Municipal Authority, Shavertown, Pennsylvania
Elrod-Dunson, Inc.
Gannett Fleming, Inc.
HNTB Architects Engineers Planners
Johnson County Unified Wastewater Districts, Johnson County, Kansas
Killam Associates Consulting Engineers
Lexington Fayette Urban County Government, Lexington, Kentucky
Lower Paxton Township, Harrisburg, Pennsylvania
Montgomery Watson Americas, Atlanta, Georgia, and Houston, Texas

New Castle County Department of Special Services, New Castle, Delaware
Oklahoma Department of Environmental Quality, Oklahoma City, Oklahoma
Ratnala & Bahl, Inc.
RJN Group, Inc.
Rock River Water Reclamation District, Rockford, Illinois
Summit County Department of Environmental Services, Akron, Ohio
U.S. Environmental Protection Agency

Federation technical staff project management was provided by J. Robert Schweinfurth.

CONTENTS

LIST OF TABLES

LIST OF FIGURES

INTRODUCTION

In the 1970s and 1980s there was an increased interest in the environment and the introduction of new regulations in the form of the Clean Water Act. This resulted in significant investments in wastewater treatment and collection facilities and expansion of sewer systems to accommodate housing and meet the goals of a clean environment. It has been estimated that the capital cost of systems is in the trillions of dollars and represent 10 to 15% of the total infrastructure of a sewered community (NCPWI, 1988).

Infiltration/inflow (I/I) problems have long been the primary focus related to public sewer lines of collection facilities, and most, if not all, of the efforts to reduce I/I have centered around the main lines.

Communities are now facing unprecedented challenges in operating and maintaining collection facility infrastructure. As the integrity of sewer systems starts to deteriorate because of a variety of factors such as old age, poor design, lack of maintenance, and budget restrictions, the systems' ability to transport wastewater to treatment facilities is causing a number of problems. A significant problem is high flows in the collection system created by excessive I/I. This I/I problem eventually comes to the attention of the general public in the form of sewer overflows, sewer backups, facility expansions, equipment failures, permit violations, higher operating costs, and higher user fees. It is now becoming clear that private building sewer connections (PBSCs) (laterals) may even be a larger part of the collection system and contribute a significant amount of I/I. Although it is recognized that public sewer mains are becoming older and are subject to I/I, PBSCs just as old as the mains are ignored and receive little or no maintenance and inspection.

Typical factors contributing to this lack of operation and maintenance are

- The number of house laterals and their connected devices are overwhelming to monitor and isolate from the main collection system.
- Officials are reluctant to become embroiled with legal issues associated with either entering private dwellings or spending public funds for maintenance of private property.
- Lack of public funds, available staff, resources, and budget monies for rehabilitation of private sewer lines.

There are a number of corrective actions that can be taken to address private sewer line installation problems, misuse of the sewer system, and public apathy (Table 1).

Because of the sheer number of PBSCs, it would not be feasible to rehabilitate the entire system but the following cost-effective corrective actions can be taken:

- Smoke testing and television inspection to identify sources of extraneous flow entering the sewer system (these methods also have the added benefit to the homeowner in that they can identify potential health and plumbing hazards).
- Development of specifications and regulations for installation of laterals to avoid repeating past mistakes.
- A regular inspection program in which the lateral is inspected before back-filling and the home is inspected for improper connections.
- Circulation of a newsletter or information bulletin advising the property owner of the adverse effects of misusing the sewer system.

- Replacement of cleanout caps on private property.

Private building sewer connections represent a large portion of the collection system, and, as illustrated in Figure 1, there are a number of illegal connections. Many illegal connections are made intentionally to protect the property owner from water damage, without considering the effect these connections have on wastewater collection and treatment facilities. Although there are existing problems with PBSC I/I that are beyond the control of property owners (i.e., broken water lines, storm sewer problems), improvements can be made by abandoning and disconnecting illegal devices (see Figure 1). For example, if we consider that a small sump pump can pump as much as 50 000 gpd (190 m^3/d) and if only 100 of these pumps are connected to the sewer system, it is easy to see why PBSCs have such a tremendous effect.

Table 1 Corrective actions.

Problem	Possible actions
Private sewer line installation problems	Creation of regulations, specifications, and ordinances dealing with lateral installations.
	Inspection and enforcement programs to oversee future installations.
Misuse of sewer system	Budgeted monies for identifying the sources of and disconnection of illegal/improper connections.
	Public outreach program that provides property owners with information that explains the effects of improper connections and benefits of eliminating PBSC I/I.
Public apathy	Program to emphasize direct benefits to property owner: if water can get into the private sewer connections, so can tree roots, which will cause blockage and additional expense to the property owner.

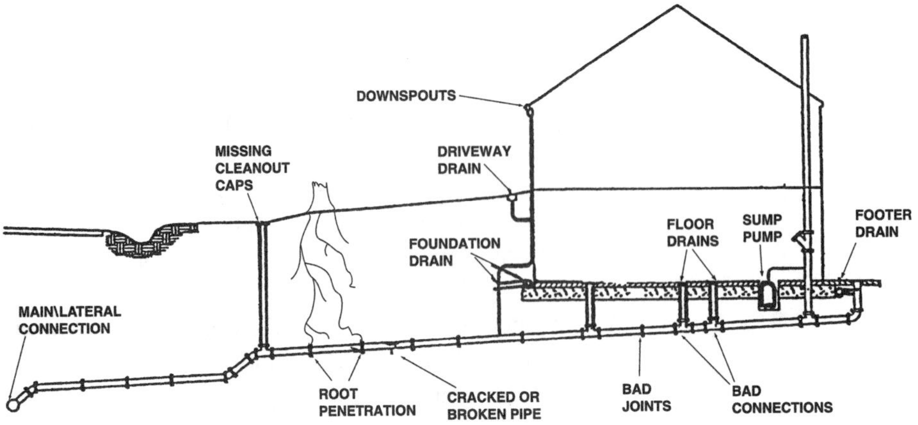

Figure 1 Typical sources of I/I from PBSCs.

EXTENT OF INFILTRATION AND INFLOW FROM PRIVATE BUILDING SEWER CONNECTIONS

Overview of the Nationwide Inflow and Infiltration Survey

Objective. A questionnaire was developed to gather information on wastewater collection systems, service connections, and the way municipalities are addressing I/I problems. A copy of this questionnaire is included as Appendix A. The questionnaire also targeted information about private property source issues, financing options, and enforcement programs.

The questionnaire was developed and peer-reviewed by the Sanitary Sewer Overflow Cooperative Agreement Workgroup, composed of members of the Water Environment Federation, U.S. Environmental Protection Agency (U.S. EPA), other professional and trade associations, and other individuals involved in wastewater collection and treatment. Findings of the survey are presented throughout this monograph.

Survey Sample Development. The questionnaire sample was developed from the National Pollutant

Discharge Elimination System (NPDES) permit holder database maintained by U.S. EPA. The NPDES permit database was chosen to represent the municipal wastewater treatment plant (WWTP) collection system population because most of the population (excluding satellite collection systems) can be found in the database. Of the 15 326 municipal treatment plant NPDES permits, 2926 of these permit holders were sent questionnaires. Additionally, 182 questionnaires were sent to members of the workgroup for distribution to owners of satellite collections systems and municipalities in their local areas not included in the NPDES permit sample.

The permit database was subdivided by the amount of wastewater the collection system transported and by respective U.S. EPA region. A computer program randomly selected permits within each subdivision. The database included subdivisions of flow by region as follows: <0.1 mgd, 0.1 to 1.0 mgd, 1 to 10 mgd, 10 to 50 mgd, and >50 mgd.

Questionnaire recipients had the option to respond anonymously. A total of 3108 questionnaires were mailed in the fall of 1997, with 316 completed (or partially completed) questionnaires returned. All percentages reported in this document have been rounded to the nearest whole number. In general, the margin of error associated with these results, based on a total of 316 responses, is +/− 5.5% at the 95% confidence level. Tests of significance have been conducted on all findings. These distinctions, when reported, are statistically discernible at the 95% level of confidence. Table 2 provides information regarding questionnaire respondents by U.S. EPA region.

Survey Respondents. A profile of respondents' ownership and administration is identified below:

- Entity ownership and administration
  - Publicly owned, 99%
    - (1) Municipal/township/ regional government, 84%

Table 2 Questionnaire respondents by U.S. EPA region.

Region	Regional office city	Total surveys sent, %	Total surveys returned, %	No. of surveys returned
1	Boston	3	3.8	12
2	New York	4	4.1	13
3	Philadelphia	11	12	38
4	Atlanta	16	12	38
5	Chicago	27	27.6	87
6	Dallas	10	8.3	26
7	Kansas City	19	20.3	64
8	Denver	5	5.7	18
9	San Francisco	2.5	2.8	9
10	Seattle	2.5	3.4	11

(2) Sewer district/board/ commission, 11%
(3) Special regional authority, 2%
(4) State or federal government, <1%
(5) Other, 1%

— Private investor owned, 1%
(1) Single sewerage system, 1%
(2) Subsidiary of larger holding company, <1%

- Entity's other administration duties
  - — Wastewater treatment and disposal, 93%
  - — Water distribution, 74%
  - — Stormwater, 38%
  - — Garbage/sanitation, 23%
  - — Other, 9%

System Description. Eighty-five percent of responding entities' systems are considered "separate sanitary" or dedicated sanitary collection systems. Eleven percent of responding entities' systems are defined as both combined and separate sanitary systems. Only 4% of responding entities' systems are combined sewer systems.

The size of the various entities responding varies greatly, with 49% indicating their entity is 5 sq mile or smaller and 17% representing an entity of more than 50 sq mile. The mean area served across the entities is 62 sq mile, while the median is 7 sq mile. For this analysis, "mean" denotes the average value reported. For example, the *mean* of values 0, 1, 2, 3, and 5 is 2.2. "Median" denotes the middle value of all numbers reported. For example, the *median* of values 0, 1, 2, 3, and 5 is 2.

In 62% of the entities, five or fewer employees are dedicated to collection system work. The mean number of dedicated employees is 21 and the median is 4.

Percentage respondents by amount of wastewater transported by collection systen is as follows:

- <0.10 mgd, 19%;
- 0.10 to 1.0 mgd, 41%;
- 1.0 to 10.0 mgd, 22%;
- 10.0 to 50.0 mgd, 12%;
- 50.0 to 100.0 mgd, 4%; and
- >100.0 mgd, 2%.

Private Builiding Sewer Connections. Information regarding percentage respondents by number of PBSCs in collection systems gathered via the questionnaire includes the following:

- <1000, 50%;
- 1000 to 10 000, 30%;
- 10 000 to 100 000, 16%;
- 100 000 to 1 000 000, 4%; and
- >1 000 000, 0%.

The median population served by survey responders is 3000. This includes 24% that service 700 or fewer and 40% that service more than 5700.

Respondents are equally divided in terms of how they define the private service lateral, with 45% indicating it runs from wye connection to building and 42% from right of way (easement) to the building.

Private sewer laterals are most frequently located at the front of the property (59%). Thirty-two percent indicated that they are located at the back of the property and 14% indicated a side location.

Table 3 lists the percent of respondents' length of separate sanitary sewer lines per number of PBSCs.

Although only 3% of respondents are responsible for maintaining private service laterals, 26% have assumed responsibility to maintain or rehabilitate the laterals. Thirty-eight percent of respondents have the jurisdiction to repair and/or replace private service laterals. Three-quarters of municipalities surveyed do not allow for multiple connections to a private service lateral.

Approximately three out of four entities require both gasketed pipe (76%) and glued pipe (74%). Gasketed pipe is less frequently required in smaller systems, that is, those with fewer than 1000 connections (66%) and those that transport fewer than 100 000 gal wastewater daily (59%). The minimum diameter pipe allowed for PBSC varies between 75 mm (3 in.)

Table 3 Percent of respondents' length of separate sanitary sewer lines per number of PBSCs.

Length of sanitary sewers, mile[a]	<1000 Connections	1000 to 10 000 Connections	10 000 to 100 000 Connections	100 000 to 1 000 000 Connections	>1 000 000 Connections
<10	29	2	—	—	—
10 to 50	19	16	<1	—	—
50 to 500	2	13	11	<1	—
500 to 1000	—	1	4	<1	—
>1000	—	—	<1	2	—

[a] mile × 1.609 = km.

(8%), 100 mm (4 in.) (82%), 150 mm (6 in.) (9%), and 200 mm (8 in.) (1%).

The vast majority (83%) of those surveyed have inspection programs in place for new PBSCs. Among those with inspection programs, the majority have been in place for at least 10 years (55%). Entities with smaller systems tend to have newer inspection programs. On average, three inspectors are assigned to this task, although the median number of inspectors is 1. Although there are sufficiently large numbers of new inspection programs, only 24% have programs that inspect *existing* private connections. This may be in part because only 5% require inspection when ownership changes hands.

Forty-seven percent of those surveyed require special training or permitting for contractors who perform sewer taps.

Sixty-three percent of entities responding accurately depict service connections on documents such as system maps, drawings, construction permits, and the like.

Of the total survey respondents, one-third could not indicate the model plumbing code used by their entity. Of those respondents who could indicate which model plumbing code is used by their entity, 45% indicate that the Universal Plumbing Code (as prepared by the International Association of Plumbing and Mechanical Officials, Walnut, California) is used. Fourteen percent of respondents use Southern States Standard as a model code, and the remaining respondents (41%) use other codes such as the entities' state codes, local codes, or site-specific engineering specifications.

Infiltration and Inflow Problems. Information gathered regarding the surveyed entities' extent of I/I experienced in their systems is as follows:

- Percentage of I/I in total flow during "average/normal" year
  - 0 to 20, 39%;

— 20 to 40, 39%;
— 40 to 60, 16%;
— 60 to 80, 4%; and
— 80 to 100, 1%.

- Percentage of total I/I related to PBSCs in "average/normal" year
  — <5, 10%;
  — 5 to 20, 24%;
  — 20 to 50, 19%;
  — 50 to 75, 12%;
  — >75, 5%; and
  — Unknown, 31%.

Only 29% of the respondents indicated that these percentages are based on monitored data. In other words, a sizeable number (31%) are unsure as to the extent to which PBSCs contribute to total I/I. This is not surprising given that 71% do not have monitored data on this measure.

Surface water fluctuations have an effect on the I/I detected in 71% of the respondents' systems. Similarly, groundwater fluctuations affect 70% of the I/I levels detected (65% of the entities base this percentage on monitored information).

Groundwater is above the crown of most service connections infrequently 43% of the time, seasonally 53% of the time, and most of the time for only 5% of those surveyed.

Only 52% of those surveyed indicate that foundation drains that are directly tied to separate sanitary sewers contribute to wet weather problems experienced by their systems.

Twenty-two percent of the entities surveyed provide check valves on private service laterals that experience backups. This is particularly true among entities with more than 10 000 service connections (43%).

Financing and Enforcement. Approximately one-half of survey respondents rehabilitate PBSCs. Among those who do, 72% of the respondents indicate that the customers pay the bill. For entities with fewer than 1000 connections, 84% of the respondents indicate that customers pay the full cost of the repairs. Other financing sources used to pay PBSC repairs include user fees and local funds. Entities for whom the customer pays for rehabilitation most commonly arrange for the customer to pay the full cost of repairs (76%). Only 20% of respondents have payment plans available. Only 39% of respondents indicate that they have taken measures to force or motivate customers to make repairs.

The vast majority of entities surveyed (83%) have enforceable regulations prohibiting downspout, roof drain, and area drain connections. These entities most frequently impose fines (23%) and disconnect water service (23%) to enforce disconnection of illicit lines. Additionally, 15% say their entity does nothing to enforce disconnection. Only 6% of respondents allow downspout, roof drain, or area drain connections.

For 59% of respondents, it is legal for one private sewer lateral to traverse another property. In those cases, 89% of the entities require easements.

Information Management Systems. Only 39% of those responding use computer technology to track facility information. Among those using this technology, flow data (44%) and preventive maintenance (25%) are most frequently tracked. Slightly more than one-third (35%) of the entities track I/I evaluations. Areas with 1000 or more connections are more likely to monitor such information.

Public Relations. Table 4 provides information regarding public relations and education programs gathered from responding entities.

Literature Review

Concern about the effect of PBSCs has been expressed since the construction of civilization's first sanitary sewer system. According to Gray (1940), the ancient Romans did not allow private households to connect

Table 4 Respondents' public relations and education programs.

Program component	Amount of wastewater transported by collection system						
	<0.1 mgd,[a] %	0.1 to 1.0 mgd, %	1.0 to 10 mgd, %	10 to 50 mgd, %	50 to 100 mgd, %	>100 mgd, %	Overall, %
Public relations program in place	24	21	20	35	25	86	25
Employee trained in public relations	28	29	38	29	42	71	33
Public education materials related to PBSCs	23	17	30	45	33	29	25
Public awareness programs	38	39	39	33	18	71	38
Procedures to receive and evaluate complaints	40	52	71	76	58	100	58

[a] mgd × 3785 = m^3/d.

to the great sewers of Rome. Household liquid wastes entered the Roman system through street gutters. In modern sewerage practice, the important role of PBSCs has also been recognized. Metcalf and Eddy (1930) noted that many engineers advocated that municipalities assume full responsibility for construction of PBSCs to ensure the integrity of the system. However, Metcalf and Eddy conceded that good plumbing codes, properly enforced, would also yield high-quality PBSCs.

Reported Problems/Magnitude. Infiltration and inflow to municipal sanitary sewer systems caused by PBSCs were recognized in the early 1960s (Van Natta, 1963). At this time, the City of Fort Worth, Texas, determined that poorly constructed PBSCs were the primary cause of infiltration to the city's sanitary sewer system. Other authors noted the presence of I/I from laterals and connections at the main line sewer but did not assess quantities or effects of I/I from building laterals (Clapham, 1965; Karanik and Johnson, 1976; Lenahan and Herdon, 1972; and McCarty, 1973).

The Construction Grants Program of the 1970s and 1980s required that sewer system evaluation surveys (SSESs) be conducted. After SSESs were performed, better assessments of the magnitude of I/I resulting from PBSCs began to appear in the literature. Roberts (1979) reported that defective service connections contributed less than 0.5% of excess flows to the Waukegan, Illinois, wastewater system. However, private source inflow contributed approximately 50% of total measured excess flows.

In a study for U.S. EPA, Conklin (1981) noted that many sewer rehabilitation programs eliminated approximately 0 to 30% of I/I, despite engineer predictions of 60 to 90% I/I removal. Building connections and nonrehabilitated pipe joints were the primary sources of remaining I/I in rehabilitated sewer systems, with private source I/I contributing 50% of the total system I/I.

Hannan et al. (1996) reported that the Washington Suburban Sanitary Commission (Hyattsville, Maryland) determined that 43% of I/I originated from sources other than the sewer main. In addition, for I/I attributable to PBSCs, 24% came through service lines and 76% entered through the connection at the main line. Curtis and Krutsch (1993), in a study of the Unified Sewerage Agency of Washington County (Oregon) system, noted that service connections contribute 50% of extraneous flows.

Representative Correction Programs. Basic methodologies for removing or eliminating I/I from PBSCs were typically noted in the literature concerning the private source I/I problem. Van Natta (1963) identified revision of the Fort Worth plumbing code as the best method for eliminating further contributions of I/I through service connections. Unearthing and repairing service connections was mentioned since the 1960s as a highly effective repair method (Clapham, 1965; Curtis and Krutsch, 1993; Kipp, 1983; Lenahan and Herdon, 1972; and Paulson et al., 1984). In addition, methods such as lateral grouting (Clapham, 1965) and lining (Steketee, 1981) were also mentioned. Progress toward developing microtunneling machines capable of installing 4- and 6-in. service connections, including connection to the main line, was reported by Fedotoff et al. (1991).

Determining who was responsible for the repair of defective PBSCs was often cited as the primary hindrance to implementation of service connection repair programs. Steketee and Beck (1983) observed that dividing ownership and construction responsibility for services (between city and homeowner), properly locating service connections, and gaining access to property were significant problems in developing a comprehensive service connection rehabilitation program for Castle Rock, Oregon.

Paulson et al. (1984) described similar problems in developing a program to eliminate private source inflow in Normal, Illinois. The town of Normal found it impractical to hold homeowners responsible for rehabilitating service connections to fix an I/I problem that had no direct negative effect on the homeowner. The town eventually began the repair work until it gained sufficient knowledge of how to repair laterals and remove private inflow sources. The town eventually assumed full financial responsibility for replacing service connections (except for landscape development), with homeowners choosing contractors from a list developed by the town.

More than a decade later, Curtis and Krutsch (1995) reported similar problems in developing a comprehensive service connection rehabilitation program for Washington County, Oregon. Gaining access to private property was an ongoing problem. They reported a 95% participation rate after the Unified Sewage Authority of Washington County changed its cost-sharing burden from 50% responsibility to 100% responsibility for PBSC repair or replacement costs.

Range of Corrective Costs. Costs for rehabilitation of PBSCs were also reported in the literature. Costs are difficult to interpret because of the effect of inflation in the 1970s and the widely varying factors associated with construction costs. Roberts (1979) reported a cost of $600 for repair of defective service connections, with a projected cost of $320/m^3·d ($1200/1000 gpd) I/I removed. This was the highest per gallon cost of I/I removal for all of the rehabilitation costs in their study. Paulson et al. (1984) reported a cost of $1500 per household for excavation and repair of PBSCs in Normal, Illinois. In 1984, the Washington Suburban Sanitary Commission reported a cost of $500 to $600 per sewer service connection for grouting versus an estimated cost of $3000 to $3600 for excavation and repair or replacement. Carter et al. (1986) provided an estimate of $1200 to $4000 for replacement of leaking service connections. However, with the extensive use of trenchless technologies, the cost of using this new rehabilitation technology is becoming more competitive.

CONTROL OF INFILTRATION AND INFLOW

Overview of Operation and Maintenance Practices

Throughout the United States, operation and maintenance (O&M) practices of responsible wastewater collection system owners are fundamentally similar, but each likely has some system-specific needs. These may require the commitment of unique employee skills, maintenance approaches, and other resources, resulting in atypical components to the organizational structure.

Typical Organization Structure and Staffing. The following case studies briefly describe the O&M and I/I control responsibilities of representative large, medium, and small wastewater collection system agencies, respectively.

LARGE-SIZED AGENCY: CITY OF HOUSTON, TEXAS. The City of Houston serves 415 000 residential, commercial, and industrial customers distributed over a 1600-km^2 (617-sq mile) service area. Wastewater is conveyed to one of 46 WWTPs.

The Department of Public Works and Engineering, Utility Maintenance Division, has the responsibility for operating and maintaining the 9000-km (5600-mile) sanitary sewer collection system. The system consists of main line sewers ranging in size from 150 to 3600 mm (6 to 144 in.) in diameter. Services are 4, 6, or 8 in. in diameter.

The Utility Maintenance Division employs a staff of fewer than 700 to manage and maintain both the water distribution system and the gravity sanitary sewer collection system. Approximately 45% of the staff are responsible for maintenance of the sanitary sewer system.

OPERATION AND MAINTENANCE PROGRAM. The Utility Maintenance Division undertook a $65 million citywide physical inspection program from 1989 to 1992. This work identified the need to rehabilitate approximately 25% of the collection system. Smoke testing conducted as part of this work identified numerous problems on PBSCs. Repairs were made to the portion of the building service line located in the city's right of way and easements to reduce the amount of I/I entering the sewer system. Officially, installation and maintenance of the lateral/service from the building drain to the main line connection is the responsibility of the property owner. The city does not repair or clear stoppages on the portion of the services on private property, which is typically the position of large municipalities.

RESPONSIBILITY FOR PRIVATE CONNECTIONS. The majority of requests for service from Houston's wastewater customers are for sewer stoppages. The division's operating procedure calls for an investigator to respond to the customer's complaint within 2 hours of the work order issuance. If the investigator determines that a problem exists and is the city's responsibility to correct, a crew is dispatched. Each crew is equipped with a trailer-mounted sewer jet to facilitate an immediate response if a stoppage is found. If the stoppage cannot be removed by the trailer jet, either a truck-mounted jet or combination cleaner is routed to the job site. If wastewater is overflowing on the customer's property and it appears that the problem is located in the customer's service line, the crew will attempt to clear the line. This is done to minimize potential public health effects.

MEDIUM-SIZED AGENCY: JOHNSON COUNTY WASTEWATER, KANSAS. Johnson County Wastewater operates and maintains a 2400-km (1500-mile) sanitary sewer collection system and wastewater treatment facilities serving Johnson County, Kansas, in the Greater Kansas City area. The total population served is 280 000, but of particular interest are three of the oldest watersheds encompassing a population of 93 000. That area, which represents one-half of the total service area, has been the subject of an aggressive inflow control project for the past 14 years, with significant results.

The project area is composed of 1195 km (743 miles) of sewers, 17 768 manholes, and approximately 51 700 structures. The system is more than 50 years old in the oldest areas and is primarily vitrified clay pipe and brick manholes. The area is dominated by single family residential homes, with associated commercial properties and some light industry.

OPERATION AND MAINTENANCE PROGRAM. Before conducting SSESs and implementing subsequent system improvements, the O&M staff spent the majority of available resources on addressing emergency calls. Routine inspection and scheduled maintenance were not a priority, resulting in numerous customer problems. Following a few intense storms in the early 1980s and resulting basement flooding and overflows, the Johnson County Board of Commissioners committed to a cost-effective solution of capital expenditures and maintenance, including construction of relief lines, sewer rehabilitation, and efforts to remove I/I from private property sources to achieve 10-year frequency design storm protection.

The private property I/I removal project consisted of 11 geographic phases, requiring that every building be inspected and tested to identify for the property owner sources of rainwater or groundwater entering the sanitary sewer system. Reimbursements to property owners are made at published rates for elimination of foundation drain sump pits, storm sump pumps or pits, area drains (driveway, patio, yard, window well, and basement entry), downspouts, and defective service line cleanouts.

To date, approximately 15 700 sources of I/I have been disconnected from private property. Postrehabilitation flow monitoring has revealed that goals for I/I reduction in each of the 11 phases of the project have been exceeded. Johnson County Wastewater continues its I/I control efforts with a full-time staff, including a collection systems engineer, I/I planning coordinator, and four inspectors who continue to investigate the sewer system in problem areas. A routine preventive maintenance program that eliminates potential sources of problems has been implemented, including a "term and supply" contract with a cured-in-place pipe (CIPP) vendor and manhole rehabilitation contractor to make repairs to pipes and manholes, as needed.

RESPONSIBILITY FOR PRIVATE CONNECTIONS. Johnson County Wastewater owns and maintains the main line only. For the property owner, ownership and responsibility for maintenance of the service lateral begins at the point of connection of the lateral at the main to the building drain immediately outside the structure. Johnson County Wastewater has not repaired any laterals as part of its capital sewer improvements and maintenance activities and has no plan to do so in the future.

SMALL-SIZED AGENCY: DALLAS AREA MUNICIPAL AUTHORITY, PENNSYLVANIA. The Dallas Area Municipal Authority is a small system located in northeast Pennsylvania, providing sewer services for 6000 customers. Five municipalities are serviced, ranging in size from 225 to 2400 customers.

The service area primarily contains residential customers, with only 20 commercial accounts (e.g., car washes, laundromats). In addition, because portions of the area are rural, the authority is responsible for permitting, testing, and inspecting on-site septic systems in three of the municipalities.

The sewer system has approximately 190 km (120 mile) of gravity sewer line ranging in size from 150 to 750 mm (6 to 30 in.) in diameter, approximately 20 km (10 mile) of force main pipe from 100 to 350 mm (4 to 14 in.) in diameter, and 10 pumping stations with capacities to serve as few as 25 homes to the main pumping station of 2.84×10^4 m^3/d (7.5 mgd) capacity. Three operators are responsible for operating, maintaining, and inspecting the sewer system.

OPERATION AND MAINTENANCE PROGRAM. This authority, like many small systems, had little equipment for maintaining the sewer system and, in fact, clearing blockages in the main was sometimes the only attention a sewer line would receive for years at a time. Realizing that this situation was not responsive to customer needs, efforts were made to have equipment available that would minimally handle emergencies such as blockages. Monies were set aside every year to acquire collection system equipment, and a used sewer rodding machine and a small trailer-mounted water jet were purchased from a larger municipal system. This equipment was used for emergencies and maintenance of some troublesome lines and, time permitting, operators cleaned sewer lines as part of a preventive maintenance program. In 1995, the used equipment was traded in for a combination high-pressure water jet cleaner and debris-vacuuming truck, which is used in maintaining the sewer system and various pumping station wet wells.

RESPONSIBILITY FOR PRIVATE CONNECTIONS. The authority subscribes to a telephone-answering service, which takes emergency calls after hours, on weekends, and on holidays. When an emergency call is received, a system operator is paged and investigates the alleged problem by going to the customer's home and checking upstream and downstream manholes for flow problems. If the problem is in the main, the Jet/Vac is used to correct the problem. If the problem is found to be in the homeowner's service line, a small hand-carried water jet with 30 m (100 ft) of 8-mm (0.25-in.) pressure line can be inserted to the plumbing of the home at the building drain cleanout to clear the line.

A simple complaint form is kept for each customer complaint. These forms are used to identify and track problematic sewer lines and provide a basic record of the incident if there is a subsequent claim for damages because of a sewer backup.

Design Misapplications and Problems

Private services, including their lateral connection to the public sewer main line, often have defects that contribute varying quantities of I/I to the wastewater collection system. Many of these defects are attributable to poor design and improper construction. These problems are found in both newly constructed and rehabilitated sewers. The "dissection" of the service line portion of the wastewater collection system into its component parts illustrates the more common problems and their solutions. They are summarized in Table 5.

Connection to the Main Line. Service lines and building sewers account for a significant footage of pipe relative to the total length of pipe in a collection system. One manhole-to-manhole section of main line several hundred feet in length may have several times this footage in service and lateral lines connected to it. Depending on the condition, this additional footage may contribute significant amounts of I/I.

The connection could also impose a significant structural load on the sewer main where connections require a stack or riser. The amount of main line pipe removed and any additional damage to the pipe barrel as a result of a plumber's or contractor's break-in or tapped-in service also weaken the pipe and make it less capable of supporting the load. Improper tapping techniques also can result in leaks. Finally, improper bedding or disturbed bedding under the joint of pipe where the service connection is made could increase the failure potential. A large percentage of the total number of repairs completed annually by collection system maintenance staff are either failures of the service near the connection to the main or the main itself where the connection is made. Each of these defects could contribute substantial quantities of I/I to the system.

To reduce failures, factory fittings (tees or wyes) should be used for all service connections to new sewers or repairs to existing sewers. If plastic pipe was used to construct these new sewers or to repair an existing sewer, a gasketed, full-body fitting should be used. For rehabilitated sewers that are sliplined or pipeburst, use of a gasketed polyvinyl chloride (PVC) saddle for service restoration is relatively common. A recent demonstration project in Houston, Texas, has shown that a bead of pure silicon caulk on the top and bottom surfaces of the elastomeric gasket between high density polyethylene (HDPE) pipe and the PVC saddle allows the connection to pass a low-pressure air test, indicative of watertightness.

Two other products for relatively watertight service connections warrant consideration. The Inserta-Tee (Inserta Fittings Co., Forest Grove, Oregon) involves a precise hole being cut in the HDPE main line pipe; a grooved, donut-shaped gasket is placed around the edges of the hole; and a short piece of PVC pipe with a bell is tightly fitted in the gasket. The "fused" connection involves a hole augered down to the top of the HDPE pipe; a riser or stack of HDPE pipe is fused or butt welded to the HDPE main line at the proposed tap location; the fused connection is air tested; and finally the hole is cut into the HDPE main to restore the service.

When sanitary sewers are rehabilitated by the CIPP method, alternatives to remote service restoration, in which a hole is internally cut remotely to restore the old service connection to the new CIPP liner, should be

Table 5 Summary of common lateral/service defects and possible solutions.

Part of lateral/service	Problem	Solution
Connection to the main	• Damage to the main line from plumbers' "break-in" tap • Improper bedding or "soft spot" under the service connection • Load imposed on the main line by the service (particularly when a stack is included on deep sewers) • Crushed or disconnected services by adjacent construction • Use of improper tapping method • Use of noncompatible materials	• Reconnect by excavation, if possible, using a gasketed polyvinyl chloride saddle with a bead of pure silicon on the top and bottom surfaces of the elastomeric saddle gasket for both new sewers or repairs to existing sewers. • For CIPP pipe, the reconnection can be enhanced by moving (relocating) the service 1 ft or so upstream or downstream of the original location.[a] • Connect service to main line at 15 to 45° from springline. • Use proper class pipe for soil and loading conditions. • Closed-circuit television test existing services adjacent to new underground construction in right of way.
Pipe joint	• I/I • Root intrusion • Faulty or incomplete joining (e.g., missing gaskets, no glue)	• Use elastomeric gasket or glued joint for better sealing. • Use thermoplastic pipe with longer pipe lengths to reduce number of joints, • Perform air or hydrostatic testing.
Pipe bedding	• Improper support for both rigid and flexible pipes	• Pipe should be supported uniformally. • For wet trenches, use well-graded small rock or shell. • Flexible pipe must be backfilled to 6 to 12 in. above pipe[b] with selective material to provide the required side support.

(continued)

Table 5 Summary of common lateral/service defects and possible solutions (*continued*).

Part of lateral/service	Problem	Solution
Cleanout	• Unable to determine if stoppage is located in service or lateral (i.e., whether a municipality or a property owner problem) • Damage to cleanout during final site grading and landscaping	• Install a cleanout at the property line to enable verification of stoppage location and serve as an access for service (only applicable where property owner is *not* responsible for the service from the building drain to the main line connection; may be difficult to retrofit existing services). • Reinspect cleanout/riser after final site grading.
Pipe	• Improperly installed (e.g., flat grade) • Debris inside line • Circumferential and longitudinal cracking • Material susceptible to sulfide corrosion; noncompatible materials	• Improve grade control. • Clear debris from inside pipe while installing. • Use proper class pipe for soil and loading conditions. • Use new, proven plastic materials.

[a]ft × 0.304 8 = m.
[b]in. × 25.40 = mm.

considered. Remote reconnects could allow much of the I/I that is eliminated from the main line by the continuous, jointless CIPP method to enter through the service connection recently restored by cutting a hole in the CIPP pipe at the location of the original service connection. Some users of remote reconnects have had success with grouting the reconnected service to get the desired watertightness, but at an added cost. Methods for remotely relining the service connections have been developed by some CIPP vendors and contractors, but these methods are currently costly. In many locations, bid prices for restoring an existing service using the remotely cut hole and the price of a new service reconnection by excavation are comparable.

Recent construction experience has shown that the quality of service reconnections to CIPP pipe by excavation may be significantly enhanced by moving the location of the actual service connection approximately 1 ft upstream or downstream. The CIPP method of liner insertion and curing tends to cause the CIPP pipe to "balloon" at original service connections. This provides a pipe that may be irregular and larger in diameter than the original pipe. It is difficult, if not impossible, to get a gasketed PVC saddle to seal to this irregularity.

Although unlined, unreinforced concrete pipe is not typically specified for new installations, there is plenty of it underground in the United States to which new connections and reconnections are being made. However, there are liabilities to making reconnections and new connections to deteriorated concrete pipe. Questions arise as to whether the deteriorated pipe will be able to structurally support the service. Another issue is whether the service installation method (e.g., nonfactory tap) could result in fracturing the concrete pipe and/or protruding the service into the main line. Increased inspection and informed plumbing and contracting industries can control exposure to these problems, but rehabilitation before any service connections are made should be considered.

Faulty Pipe Joints. Improper joining of pipe sections is another potential source of I/I. The Uniform Plumbing Code allows for use of solvent welding of PVC pipe joints for PBSCs. These lines are typically shallow and have relatively little opportunity for infiltration if installed and maintained properly, unless high groundwater conditions exist. The deeper portion of the service line located in the easement or street right of way, that is, the lateral, is more susceptible to infiltration. The lateral may also be a larger-diameter pipe than the building sewer or service, possibly a 150-mm (6-in.) instead of a 100-mm (4-in.) pipe.

Because of the deeper depth and larger diameter of the lateral, use of an elastomeric gasketed, thermoplastic pipe for lateral replacement should be considered. Plastic pipe also allows for longer pipe joint length than other materials, reducing the total number of gasketed joints—fewer joints, less opportunity for I/I to enter the pipe. However, there is no standard for the profile of the elastomeric gasket in the pipe bell. Some gasket shapes seem designed to provide squeegee action to clean the pipe spigot before it is pushed home into the bell, preventing debris from disrupting the gasket seal. Other gasket profiles may reduce the opportunity for the gasket to be displaced, or "rolled," in the bell when the joint is made. Debris on the spigot or a rolled gasket could inhibit a tight seal.

Pipe Bedding and Backfill. Another consideration is the pipe bedding design. All pipe, but especially flexible pipe, should be properly bedded and backfilled. A rigid pipe is designed to withstand a specific load. With flexible pipe, the imposed load is designed to be supported by the pipe and adjacent soil or fill material. Pipe bedding, whether the pipe is flexible or rigid, should support the pipe uniformly. For wet trenches, a well-graded, small rock or shell (oyster or clam) works well. It also provides firm footing for any workers in the trench. In dry trenches, the bedding required may be just smooth, undisturbed, consolidated native soil; a clean compacted sand; or compacted cement-stabilized sand.

The other component of flexible pipe installation is the initial backfill placed around the sides of the pipe to 150 to 300 mm (6 to 12 in.) above the pipe. This material may be a well-graded small rock, gravel, coarse sand, or shell. The initial backfill must provide support to the sides of the pipe to prevent short-term pipe deflection during the completion of backfilling and long-term pipe deflection with both the consolidation of the backfill over time and any postconstruction loads.

Cleanouts. Another integral part of the PBSC is a cleanout installed just within the right of way or easement limit. It serves as a point of access for servicing the sewer and as a "port" for visually inspecting or televising the sewer and determining whether a stoppage or blockage is occurring downstream. If the cleanout is surcharged, a full or partial blockage exists downstream. If a sewer backup occurred in the building and the cleanout is flowing or no flow is visible, the blockage or stoppage is located between the cleanout and the affected plumbing fixture(s). At this time, the cleanout becomes a point of access where various sewer cleaning tools may be introduced to the sewer to remove or dislodge an upstream or downstream blockage or debris.

The location of the cleanout is critical. Plumbing codes typically require a cleanout at changes in direction of the sewer or where sewer cleaning tools may become snagged, unable to negotiate the turn or bend. They are also placed approximately every 30 m (100 ft) of pipe run, depending on local requirements, such that the standard length for sewer cleaning equipment is not exceeded. Many municipalities require a cleanout at the property line bordering the easement or right of way. A cleanout in this location would allow a plumber or city maintenance personnel to determine if a problem is in the city-maintained or private portion of the service/lateral, introduce the tools required to clear the stoppage, and correct the problem. This is assuming that the municipality has or takes any responsibility for the service from the building drain to the main line connection.

Pipe Material. In the past, most services were either clay or concrete pipe. Pipe joints were either made with an asphaltic or bituminous material or with cement mortar. Over time, clay pipe typically cracks both circumferentially and longitudinally because of installation practices and subsequent loading, with the longitudinal cracking occurring at the crown (top) and spring line (one-half the diameter length below the crown) of the pipe. When this occurs, the pipe is said to have "quartered" and deflected vertically. Interestingly, as the sides of the pipe move outward and meet the resistance of the soil around the pipe, the pipe again becomes stable. Concrete pipe typically fails over a period of time as a result of sulfide corrosion. Some plastic pipe sewers have both design and installation problems as previously described.

One of the more significant pipe problems occurs with the reinforced plastic mortar (RPM) pipe, or truss pipe. This pipe consists of a thin outer plastic tube and a thin inner plastic tube separated by a plastic truss network. A light-weight cement mortar is placed in the truss network as a fill material that provides additional stiffness. Sections of pipe are connected by a solvent cement coupling. Services are connected with a hole being cut into the pipe and an ungasketed saddle being solvent cement welded to the pipe barrel. Finally, a 150-mm (6-in.) solid wall acrylonitrile butadiene styrene (ABS) or PVC pipe is solvent welded to the coupling on the saddle.

The more critical problems noted with RPM pipe have been improper service connections and pipe collapse. The "holes" tapped into the pipe for service connections are often not properly cut but broken in with a hammer or cut square if the plumber did not have a key hole saw and used a hacksaw or a handsaw. This abuse results in severe damage to the pipe. The unsealed cut-ins and any tears or punctures in the thin plastic tube result in groundwater infiltration or wastewater exfiltration through the cementitious grout-filled truss network. Over time, the grout is depleted, possibly by erosion or corrosion, with a loss of pipe stiffness resulting in possible pipe failure. Attempts to seal coupled joints by injecting an acrylamide grout have not always been successful, with the gel grout migrating through the truss network to the next pipe joint or service cut-out.

Diagnostic Techniques for Private Building Service Connection Infiltration and Inflow Detection—Standard Infiltration and Inflow Sewer System Evaluation Survey Procedures

Smoke Testing, Dyed Water Flood Testing, Air Pressure Testing, and Closed-Circuit Television Inspections. Smoke testing is a relatively inexpensive and quick way to detect I/I sources from laterals and services. Positive smoke tests followed by confirming dyed water flood testing allow further fixing of the sources' locations. However, smoke testing has its limitations. In areas where groundwater is seasonally or consistently high and laterals are below the groundwater elevation, smoke testing has little or no application. Furthermore, lateral infiltration sources under sidewalks, road pavement, and other impervious areas may not be detectable with smoke testing or may produce positive smoke tests away from the actual source because of smoke migration to a soil fissure or pavement defect away from the actual source. Finally, the overall repeatability of smoke test leaks could be less than 50%, possibly causing a municipality to be resistant to take enforcement action against a property owner. A recent rehabilitation demonstration project in Houston, Texas, identified an overall 25% repeatability rate of smoke test results in two sewer basins, each with 6000 m (20 000 lin ft) of main line sewer, when such data acquired in 1994 were contrasted with those acquired 2 to 4 years earlier.

Dyed water flood testing involving the use of fluorescent dyes and large quantities of water to confirm positive smoke tests on laterals is a useful practice. The procedure requires the dyed water to be applied to an area, isolated by sand bags or other measures, that exhibits the smoke emission. The presence or absence of dye in the immediate downstream manhole indicates the existence of infiltration. Closed-circuit television (CCTV) used in conjunction with dyed water flood testing allows further pinpointing of the infiltration source to a specific point where the defective lateral connects to the main line.

Low-pressure air testing is a technique typically used to ascertain the watertightness of new and rehabilitated sewer main lines and services. The test could be applied to individual existing laterals to assess the existence of infiltration sources, but other testing methods are likely less costly and produce results with acceptable quality and repeatability. The procedure includes plugging a main line section's ends at upstream and downstream manholes and the laterals at the property line or the building drain and pressurizing the isolated system to a specified pressure. Loss of air pressure must not exceed an allowable rate over a fixed time period that depends on the total length and diameter of the pipe. Should the system not hold the pressure adequately, troubleshooting the source of leaks will be an added cost.

With the advent of more affordable and easy-to-use miniaturized CCTV equipment such as the "cable" camera, which can be inserted to lines 4 in. in diameter on the end of a fiberglass rod, CCTV could become an increasingly favored tool in the arsenal of SSES equipment. This is especially true when lower cost, visual identification of defects is desirable. Although completely substituting cable cameras for smoke testing may be neither practical nor preferred, this technology could be useful in providing visual confirmation of smoke testing efforts. Additionally, in the case of inflow sources such as area and driveway drains with water traps, CCTV could become an increasingly common physical inspection tool in confirming these sources.

Flow Monitoring. Flow monitoring of laterals and services with diameters of 150 mm (6 in.) or smaller is a difficult and more expensive approach to identifying I/I from private property sources. Typically, such data collection depends on manual recording or videotaping of weir measurements, because automated equipment is not readily available for this application. Furthermore, the smaller the tributary area to a measuring device, the less accurate are the measurements inherent to the highly variable and extremely low flows that could be encountered.

Age and General Condition of Sewers. The average age of sewers in a given area could indicate the probability of whether a high quantity of defects in laterals and services may be found. In many older communities and neighborhoods, materials of choice for pipe and related appurtenances are now known to be less durable than previously understood. Experienced service lives have fallen well short of what manufacturers represented. Such has been the case with unlined and unreinforced concrete pipes and earlier versions of reinforced plastic, mortar-lined pipe. Jointing materials used in older pipes have proven to be highly susceptible to degradation with time, leading to leaking joints. Additionally, the workmanship of the local construction contracting community and the level of inspection by the municipality could have markedly improved over time. Thus, older construction should be inherently more suspect of having I/I-related problems until proven otherwise.

Finally, a sewered community could have been annexed by an adjacent city with more stringent sewer ordinances, plumbing codes, code enforcement, and sewer design standards aimed at minimizing I/I-causing defects. That community should be scrutinized for the possible existence of a relatively inordinate amount of lateral and service defects, especially if land developers and builders were responsible for connections from the main line sewer in public rights of way or easements to the building drain on private property.

Pumping Station Monitoring. Increased wet weather flows to a pumping station could indicate the presence of new I/I sources in the tributary area to the station. There are several methods for obtaining the necessary data to support evaluation of these flows. These include chemical/physical characterization of wastewater samples and hydraulic analyses. It should be understood that, unless the pumping station is serving a relatively small number of customers, such analyses will not be particularly helpful in pinpointing I/I sources, as is the case with smoke and dye water testing and internal television inspections. Nevertheless, results from these efforts could supply the first step in prioritizing and focusing subsequent investigations.

The analysis of wastewater samples for such physical and chemical constituents as total suspended solids (TSS) and chlorides can provide important data. When concentrations of TSS are contrasted with typically expected or historically measured concentrations, a consistently lower concentration may indicate an increased component of groundwater or rainwater being introduced to the system. Similarly, increasing chloride concentrations could indicate that brackish groundwater is entering the system. This information, with flow data, could support rough estimates of quantities of "clean" water entering the collection system using mass-balance computations.

Hydraulic analyses depend on the availability of adequate flow monitoring data. Historic flow data from pumping station flow recorders and pump runtime totalizers may provide all or some of the required data. However, should historic, continuous flow monitoring data not be available, several options exist for obtaining the data, including

- Drawdown ("draw-and-fill") testing for constant-speed pumping stations in which the wet well volume is known and the pump run time is measured over time, covering dry and wet weather periods of operation.
- Ultrasonic and other types of flow meters on pumping station discharge piping, which provide both instantaneous measurement of flow rates and pumped volume data.
- Flow meters in the manhole immediately downstream of a manhole receiving pumping station discharge.
- Area velocity meters in the gravity influent piping to a pumping station.

Using combinations of two of these approaches simultaneously may provide a check of one against the other.

Rehabilitation Techniques for Private Building Service Connections

The choices for repairing defective services have expanded with the development of modern trenchless methods, including robotic spot repair, lining, and grouting as alternatives to the traditional method of abandoning or digging up the old service and replacing it by an open cut. The latter method for dealing with infiltration correction certainly has its cost advantages where conditions of shallow services under predominantly turf areas are found. However, this method could quickly become cost prohibitive when dealing with relatively deep services or when high restoration costs would be encountered because of construction impact on mature trees, shrubbery, and other plant material; garden walls; elaborate fencing; driveway pavers; and concrete fish ponds and fountains.

Generally, to date, experiences with trenchless methods are neither widespread nor extensive, but at a minimum they are worth exploring through pilots and demonstration projects. The following is an overview of available service repair methods.

Removal and Replacement. Hand shoveled and mechanically excavated trenching are tried and proven methods of installing replacement services. They can be paired with either abandoning in place or physically removing the old service. A key factor to a collection system owner using the remove-and-replace method is obtaining the property owners' consent to access the property for construction.

The initial concern of the property owner will likely be whether his or her property can be restored to preconstruction condition, especially when mature vegetation may be affected. Should it be necessary to realign the new service to minimize the extent of ground restoration or avoid structures and other unmovable obstacles, the details of a sketch will have to be negotiated and included in the owner consent agreement. Cleanouts will have to be located per the local building code, including (1) immediately outside the structure at the building drain, (2) at specified angles of changes in direction, and (3) where maximum straight runs of pipe are exceeded. Those locations could be controversial if not in places desirable or convenient to the property owner. Finally, a warranty will have to be extended for materials and workmanship. Sod and other landscaping material often will suffer, resulting in claims for warranty service. Pre- and postconstruction conditions should be recorded on videotape.

Spot Repairs. The need for repair of one or more relatively short portions of a service may lend itself to use of a spot repair, either trenchless or open cut.

TRENCHLESS. Two remote trenchless methods use an expandable plug with repair materials enveloping it. The plug is pulled into place, guided by a CCTV camera. Access is typically accommodated through a manhole. Alternatively, access can be accomplished by excavating and breaking into the line, but this is less desirable. One method uses a stainless steel sleeve with grouting on the outside. It is expanded with 280 kPa (40 psi) pressure at the location requiring repair, locking it in place. The operation takes less than 1 hour on a clear line. This procedure can also serve as a structural point repair in addition to one correcting I/I sources. Another method uses an open-ended tube saturated with resin carried on the outside of an expandable plug. The assembly is moved into position, expanded, and held tightly in contact with the host pipe. A resin capable of curing at ambient temperature is used to avoid the need for supporting the process with a heating system as used in traditional sectional and manhole-to-manhole CIPP rehabilitation. Total installation time is approximately 2 to 3 hours. Both methods can support the repair of services with a nominal diameter of 150 mm (6 in.), with lengths up to 0.9 m (3 ft). Thus, their use would be precluded in instances when 100-mm-diam (4-in.-diam) services are allowed by code.

OPEN CUT. Digging up and repairing a defective section of service may be the most cost-effective approach. This will likely be the case when dealing with relatively shallow services, when the repair is not in a well-landscaped area (e.g., mature trees and bushes, fountains, and fish ponds), or where elaborate "hardscaping" (e.g., brick fencing, drives and sidewalks, swimming pools) will not be affected. Under this method, the repair site will simply be excavated, the defective section removed, and the new section of pipe installed with proper field closure to ensure watertight connections to the existing service.

Liners. Recently, experiences with lining the portion of the service lateral in public rights of way has been positive, with increasing use. There has been limited experience to date with liners on private property, although in the near future this application is expected to increase. One issue that could arise when using either a CIPP or plastic slipliner is whether any plumbing code changes will be required as a result of the reduced interior diameter of the service to something smaller than the nominal 100 mm (4 in.) typically required. This issue will have to be dealt with on a case-by-case basis.

With the use of CIPP, access to the service can be from the main or from the building cleanout. If a cleanout is not available, a small entry point outside the building can be made and one installed. The process of inverting a resin-saturated tube to the service, first cleared of roots and other debris, is followed by heat curing. This results in a structurally restored and watertight line and is similar to the application of the same process in mains.

Typically, installing a slipliner requires excavating entry points at both the upstream and downstream ends of the service for insertion of the continuous flexible liner and pulling the liner into position. Once in position, field closures at either end of the liner are made to form a watertight connection to the existing lateral and building drain. However, there are products available that access the lateral from the main line without any access from the private property side.

Pipe Bursting. Pipe bursting and pipe reaming may be attractive alternatives to lining or remove/open cut/replace and share the advantages of liner installation. Similar to installing a lining, excavated entry points at both the upstream and downstream ends of the service are required. One entry is for insertion of the end of the replacement pipe led by a static or dynamic cutting or crushing head (see Figure 2), while the other end is for access to pull the head with attached pipe into position by cable or rods. However, with pipe bursting there is no permanent use of a host pipe; the existing service is only used as a pilot hole for the new pipe and is destroyed or removed as the cutting head progresses. A clear advantage over lining is that the new pipe can be upsized a pipe size diameter or two, if necessary.

Grouting and Epoxy Injection. Chemical grouting is a common method for sealing leaking joints, small holes,

Figure 2 Building sewer pipe bursting tool and replacement pipe for a 100-mm (4-in.) line. (Courtesy of TT Technologies, Inc.)

and radial cracks in laterals and mains but has no structural enhancing qualities and may have application to services (see Figure 3). One proprietary grouting method requires the service to be internally flooded with chemical grout, with the lateral and perhaps the main, in a two-step process. After all surplus liquified grout is evacuated from the pipe, the two-component grout remaining in the cracks and the soil matrix immediately outside the pipe cure to form a watertight seal. More conventional grouting requires insertion of a packer with collars at either end, which allow the packer to isolate a leaky pipe joint. The packer is guided into position under CCTV monitoring; the joint is isolated and air tested; and, if required, a low-viscosity grout is pumped into the isolated joint, curing to form a resilient gel. The packer is then pulled to a new position and the process is repeated.

Epoxy injection, although similar to grouting, provides the added benefit of returning some structural integrity

Figure 3 CUES sealing system setup. (Courtesy of TRB Specialty Rehabilitation, Gambrills, Maryland.)

to the rehabilitated pipe. However, because epoxy is more viscous, it cannot be pumped as far as grout. This physical feature has resulted in limited applications, such as robotically rehabilitating broken main line pipe and service connections and sealing around the area where the lateral opening has been made in the liner.

Disconnection of Inflow Sources. Although disconnection of inflow sources is not a repair of the service per se, elimination of illicit sources of extraneous stormwater connected to the service is important. Primary inflow sources and their prescribed method of elimination are as follows:

- Rain gutters/downspouts—eliminate at service or main connection and redirect downspout discharge to local drainage; some regrading of the lot may be required to redirect additional stormwater flow via a shallow lot line ditch to the front lot where drainage facilities are located.
- Foundation drains/sump pumps/area drains—eliminate at service or main connection and redirect discharge to local surface drainage, possibly requiring installation of new sumps from which to pump to storm sewer extensions; an increase in local groundwater elevation could result and may exacerbate local drainage difficulties and cause new sewer infiltration problems or basement dampness or flooding.
- Damaged or open cleanouts—restore threaded cleanout hub and install new plug; replace nonsealed cleanout with a sealed cleanout; replug all open cleanouts.
- Stormwater drains cross connected to laterals—eliminate cross connections and reroute stormwater drain.

Socioeconomic Ramifications

In addition to the large number of engineering, construction, and operational issues associated with I/I in PBSCs, utility managers must also address some socioeconomic issues. Most importantly, are there disparate effects on different neighborhoods within the service area; issues related to public or private funding of improvements on private property; and potential effects on elderly or fixed-income utility customers?

Older/Lower Income Areas More Heavily Affected. First, it is generally true that private sewer laterals will tend to be most in need of repair in older neighborhoods. Laterals under older subdivisions have had more years to endure general degradation from the environmental conditions and old and inferior construction methods and materials, more chances to sustain damage from vehicles traversing the yard, more exposure to mature trees, and more opportunities for damage from various excavations.

Challenges for the utility manager result from the tendency in many communities for older neighborhoods to decline in terms of average property values and median income levels relative to other, newer neighborhoods in the same metropolitan area. By focusing rehabilitation efforts on areas with the most demonstrated need, managers could be accused of unfairly targeting poor areas in the community. The obvious and most powerful answer is to conduct a thorough job during the evaluation and targeting process so that evidence of the logical selection of neighborhoods is readily available.

Relatively lower income in a target area, alone, is an economic issue that will be addressed below. From a socioeconomic standpoint, a problem can result in low-income areas from a rehabilitation program designed to have property owners pay the costs (or most of the costs) of repairs on private property. Utilities pushing property owners to spend large dollar amounts relative to household income are likely to encounter resistance that will further compound the difficulties resulting from purely financial concerns.

Other than direct or indirect funding by the utility in some form, an option to address socioeconomic issues is through public education and involvement. It is important to invest management and staff time in community relations regarding the utility's rehabilitation efforts. Communicating the importance of lateral repair is critical to public acceptance of any rehabilitation project and is especially important when private property or private expenditures are involved. Up-front investment in community relations may seem to be only an additional overhead project expense but, in actual practice, this type of investment can make the difference between a successful project and a significant public relations failure.

Effect on Elderly/Fixed-Income Population. In addition to the challenges associated with low-income areas, utility managers must also face the special issues associated with projects in neighborhoods with significant percentages of elderly or fixed-income residents. Elderly and fixed-income residents typically represent a greater percentage of residents living in older, targeted neighborhoods as opposed to newer, nontargeted neighborhoods.

The challenge being addressed here is not related to income, because income constraints generate the same problems without regard to age or income source, and these are addressed above. Instead, the issue relates to the actual or perceived ability of elderly household members to manage their response to this type of issue, the ability of customers on fixed incomes to budget for required expenditures, and public perceptions associated with a utility placing burdensome requirements on some of its elderly or fixed-income customers.

In response, utilities should anticipate a need to provide a certain amount of "hands-on" guidance to its customers in target areas, even if only in response to the perception that elderly residents may be less capable of dealing with the logistics of a project to rehabilitate their own failed laterals. Similarly, as the financial burden is often seen by the public in an especially harsh light if it is being placed in an area with a largely elderly population, an investment in strategies to help contain the up-front and monthly burden of any requirement from residents is likely to pay off in an increased opportunity for project success.

Public Versus Private Funding for Private Property. A logical conclusion of the above discussion on challenges associated with projects in low-income or elderly/fixed-income areas would be to focus on potential advantages of having the utility pay for some or all of the costs of a rehabilitation project.

Some states make it illegal for public bodies to spend money on improvements to private property without recovering such expenditures by assessment of the owners. Such laws have the noble purpose of preventing units of local government from inappropriate expenditures for the benefit of one person or group, but these laws can also prevent appropriate expenditures of funds on private lateral projects that would benefit to the system as a whole. For a utility with the financial resources to embrace the costs of rehabilitation on private property, restrictive legislation should not be seen as an absolute barrier. Legislative relief can be pursued, which has been successfully accomplished on a limited basis around the United States.

With regard to public perception issues, spending public money on improvements to private property has the potential to generate a problem if a disgruntled ratepayer were sufficiently motivated to challenge the fairness of the utility's rate structure. The argument could be made that a rate structure that generates enough money to allow the utility to fund improvements on private property must be over-recovering from its customers. The underlying idea here would be that, even though such an expenditure might be legal, it represents a departure from normal municipal utility practice, and, as such, it could be judged to be an excessive or imprudent expenditure that should not be included in the basis for user charges.

The answer is much the same as for other socio-economic issues: do a superb job of researching, analyzing, and documenting the need for and systemwide benefits of rehabilitating PBSCs. A utility that can demonstrate a positive and believable cost-benefit ratio for a rehabilitation project will have gone a long way toward intercepting serious challenges.

Education and Politics

Education. Many owners of wastewater treatment systems have structured information programs aimed at educating the general public about their wastewater treatment facilities. These educational programs provide information about the need for capital improvements, proposed and ongoing construction projects, operating budgets, and user rates. On the other hand, wastewater collection facilities (pumping stations and pipelines) have traditionally received little attention.

Elected officials and users of the sewer system typically do not understand the extent and nature of problems that could be experienced by the collection system, especially during wet weather. Unless such problems are documented during investigation of customer complaints or for purposes of regulatory reporting, typically, operational staff are reluctant to voluntarily disseminate reports and information about where and how frequently sewer overflows may have occurred. All too often customer contact is limited to receipt of the monthly service bill or when problems, such as stoppages, blockages, overflows, line breaks, and collapses, dictate a complaint and an immediate response by maintenance staff.

Public education on the potential damage to collection systems, environmental effects, and increase to wastewater system operating costs because of faulty services and illegal inflow sources could occur through

- Sewer bill stuffers and newsletters;
- Printed and electronic media campaigns;
- Presentations to service groups, neighborhood associations, and other interested groups; and

- School ecology programs for children of all ages.

With each of these methods of disseminating information, the responsibilities of property owners to properly maintain their services should be emphasized. If users are not properly informed, then any misuses of the collection system, whether they be willful or out of negligence or ignorance, could continue. Cooperation between the local utility's policy setters and customers is a basic ingredient for correction and control of I/I entering the public sewer system from private property sources, and it starts with educating the public.

Infiltration and Inflow Correction and Conservation of Water Resources. In addition to all of the obvious benefits of preventing I/I from entering the sewer system (such as reducing the potential for sanitary sewer overflows [SSOs]), an often overlooked benefit is the potential for decreasing effects to water resources. Once rainfall is diverted to a sanitary sewer from such devices as house rain gutters, its availability for potentially recharging local aquifers and augmenting local stream flows is permanently lost. Granted, the comingled stormwater and wastewater likely find their way to downstream surface water when discharged after proper treatment and contribute to beneficial uses of that resource. However, this probably occurs many miles from where the stormwater entered the collection system. Additionally, defective sewer pipes that allow infiltration of stormwater and groundwater could allow exfiltration of wastewater. This may negatively affect the water quality of groundwater and possibly surface water, threatening their beneficial uses. Making homeowners aware of these threats to water resources may be enlightening and provide a different but more understandable slant to the far-reaching consequences of unchecked I/I.

Governmental Subsidies to Property Owners for Correcting Infiltration and Inflow. Issues surrounding the use of public funds to correct I/I sources on private property are complicated. Their resolution is complex, and the spectrum of available options, each with unique legal issues, provides confusion for elected officials. This often results in indecision and delay in dealing with the problem. These officials will have to carefully weigh their responsibilities to protect the health of the general public, environment, and ratepayer interests against their responsibilities to their constituents who are concerned with the possible or perceived violation of private property rights associated with implementing some of those options. In their deliberations, they will need to consider things such as

- State constitutional constraints prohibiting the gifting of public funds;
- Administrative code constraints;
- Internal Revenue Service codes, maintaining tax-free status of municipal bonds, and the implication of using bond principal for reimbursement of property owners;
- Required recovery of expenditures of public funds on private property improvements;
- Access to private property to test and repair services; and
- Viable sources of funds for expenditures on private property improvements.

Some decisions may take legislative relief to allow full implementation or may require adoption of new sewer codes and ordinances. Still others may require a state constitutional amendment, rendering those options particularly infeasible or highly unfavorable. Regardless, decisions will likely be challenging and not without controversy but, at the same time, will have far-reaching benefits.

Economic Issues

This section addresses some of the economic issues associated with projects to control I/I in PBSCs.

Diversion of Funds from More Visible Infrastructure Needs. Of all public infrastructure needs, rehabilitation of private sewer laterals seems to suffer to the greatest extent from the "out of sight, out of mind" challenge. Traditionally, it has been extremely difficult to create public support for expenditures on wastewater infrastructure of any type and for collection systems in particular, but private sewer laterals are viewed by the public and most elected officials as outstandingly trivial and not a public responsibility.

Changing this perception is a challenge and will require the same sort of documentation described under the discussion of socioeconomic issues, with the addition of some viewer-friendly pictures, graphs, and charts. If rehabilitation of private laterals is critical to a utility's I/I control program, the utility will be able to develop evidence of this and communicate it to elected officials and the public. Pictures are an important first step in defining the subject area and illustrate the relationship between I/I and damage or deterioration of laterals. A pie chart would be a good choice to illustrate the relationship between linear feet of sewers on the utility's side versus the estimated number of linear feet of private service laterals connected to the system. Such a chart will help communicate the importance of focusing on laterals, by documenting the fact that these relatively short lengths of pipe can add up to a significant portion of the total sewer system and a significant contribution to the SSO problem. Finally, a number of simple graphs will be important for illustrating the cost of I/I. System users and elected officials are typically responsive to illustration of the cost burden on I/I, especially when that burden is shown as increasing charges to typical residential customers.

Effects of Taking Action. *EFFECTS ON THE TAX BASE AND LOCAL ECONOMY.* Significant I/I rehabilitation projects of any type can have positive effects on the local tax base if the problem being eliminated has negatively affected property values. The best example would be correction of I/I so a moratorium on new connections could be lifted when sufficient capacity were regained. Similar effects might be seen if regular overflows to recreational lakes or streams were eliminated. A significant positive effect on a neighborhood or area would likely be seen if recurring backups into homes or businesses were eliminated.

EFFECTS ON RATES AND CHARGES. The effect on rates and charges depends on funding methods and responsibilities selected by system managers and elected officials. If it is determined that private property owners should bear 100% of the costs of rehabilitation, the only rate effect will be the potential reduction in O&M costs from reduction in extraneous flows. If the utility bears part or all of the costs of the project, needed funds must come from either rates or an external source such as tax revenue.

To whatever extent revenue from ad valorem or other taxes is available for special utility projects such as rehabilitation of private laterals, rate increases can be mitigated. However, it should be noted that such external revenue sources are not typically seen as appropriate for funding of sewer maintenance. Tax options commonly available for capital spending include special district or zone tax levies and a special purpose local option sales tax. Any of these can contribute to a utility's ability to undertake special projects without undue rate increases. Such additional funding can be appropriate for projects such as these, especially when local decision makers can point to communitywide benefits from the project.

User charges represent most of the money available for most utilities. As such, they represent the ultimate support for virtually all expenditures, whether they be for operating or capital needs. Other than outside funds or tax revenue, user charges must support all of the utility's share (if any) of the cost of lateral rehabilitation. Connection charges, impact fees, and other forms of capital fees and charges are not likely to be seen as appropriate because these charges must typically be restricted to costs associated with providing new capacity for new users.

User charges must pick up any remaining burden and, as long as the utility has a demonstrated need for the projects, these costs should be readily included in the rate base, legal constraints aside. Note that even if a utility ultimately will recover 100% of the costs from property owners, rate increases might still result from cash flow needs if the utility offers assistance in low- or no-interest financing of costs. Of course, there will be offsetting savings as extraneous flows are eliminated and operating costs decrease or as billable flows replace these extraneous flows.

Effects of Inaction. *EFFECTS ON THE TAX BASE AND LOCAL ECONOMY.* Inaction can have varying effects on the tax base and local economy, depending on the severity of the I/I problem, the places where the problem manifests itself, the condition and configuration of the sewer system as a whole, available system capacity, and the degree of regulatory interest in the utility.

Some effects are easier to identify and measure than others. To the extent that failed laterals tend to contribute to SSOs, significant economic damages can result from backups to basements or flooding from overloaded pumping stations. Direct damages can occur to both private and public property or to water bodies; additional indirect damages can occur and would include environmental costs and potential effects on local economies through actions such as beach closings.

Tax base effects of inaction are often minimal, except in cases of extremely significant degradation that might allow leakage of wastewater that could contribute to health hazards. One additional material effect can occur if I/I flows are limiting treatment capacity such that a local moratorium on future hookups to the sewer system is imposed.

EFFECTS ON RATES AND CHARGES. Inaction can negatively affect rates and charges, even though there are obviously no capital expenditures associated with taking no corrective action. The problem is the increased cost of treatment and conveyance to handle I/I. As indicated in other sections of this monograph, the problem is greater than would be indicated by the amount of I/I entering the system on an annual basis. Inflow events are particularly expensive to handle because they can strain system capacity, and providing additional conveyance or treatment capacity is typically one of the most expensive options undertaken by wastewater utilities. Additionally, the cost of emergency activities in response to overflows or backups must be borne. These costs may not be a budget line item and, therefore, may not be easily tracked, but the cost is real and must be recovered, nevertheless. The question for utility managers is to determine the point at which corrective action represents a less expensive option than continuing to provide capacity for and treat I/I and respond to overflow/backup events. The answer requires a cost-benefit analysis and will be unique to each utility. Municipal officials must be made aware that inaction can lead to higher costs later.

Effect on Low-Income Areas and the Elderly/Fixed-Income Population. Low-income areas and areas with a large proportion of elderly or fixed-income residents typically represent an additional hurdle for any utility considering placing all or part of the cost of the rehabilitation of private sewer service laterals on property owners. Separate from socio-economic issues, some customers simply do not have the ability to bear the project costs.

Utilities have two primary external financing options to attempt to address problem areas: public or private funding or low-interest loans. When these are not available or are inadequate to make the project affordable, internal financing may represent the only viable option for certain target areas. Internal financing, in which the utility uses its own fund balances, can range from direct payments by the utility, to establishment of liens on the property with the charges deferred until the property is sold, to the provision of no-interest loans to residents to fund their rehabilitation project. Another option is to temporarily waive the mandated repairs of known defects until the time when the property changes ownership or to have the utility repair the defective lateral and bill the property owner.

Regulatory Issues

Contribution to Wet Weather Overflows. For many years, the effect of leaking services and the connection of illegal inflow sources (e.g., roof and area drains) to the house service or directly to the main were considered insignificant in contrast to defective main lines and manhole covers, frames, and inlet/outlet sewers. The advent of numerous SSESs, intensive flow monitoring, and physical inspections of sewers around the country, starting in the early 1970s with the reauthorization of the Clean Water Act, brought new attention to private property sources of I/I.

In the past 10 or more years, numerous technical papers have been presented and published referencing the relative magnitude of the problem. Two particular agencies, namely, the Stege Sanitary District, El Cerrito, California, and the Unified Sewerage District (USD), Washington County, Oregon, have attributed defective house services to 30 to 35% of the rainfall-dependent I/I peak rate experienced by their respective agencies. Stege found it cost effective to replace approximately 150 services in one sewer basin instead of upsizing downstream conveyance facilities, while USD is cognizant of the contribution of service defects to its I/I problem but has not committed to a repair/replacement program yet.

The Vallejo, California, Sanitation and Flood Control District conducted an SSES, with U.S. EPA concluding that the agency was experiencing extraordinary amounts of I/I in its system. To receive grant funds under Section 201 of the Clean Water Act for other wastewater system improvements, the district is committed to spending more than $600 000/yr through the year 2008 to replace every defective service. These funds are generated from user fees.

The common conclusion is that I/I from private property is significant, and some utilities are finding it cost effective to correct the problem to avoid or minimize overflows and reduce capital and operating costs of their collection systems.

Legal Authority To Correct Private Property Problems. To the numerous municipalities and special districts that have received Clean Water Act grants, one of the covenants of their grant agreement required that they implement a U.S. EPA-approvable sewer use ordinance, which among other provisions required elimination of illegal inflow connections. Even those municipalities that did not receive such grants have taken it upon themselves to voluntarily adopt such ordinances as a prudent business practice, or state or federal wastewater discharge permits have required correction of such sources of I/I.

As part of implementing such an ordinance, municipalities adopted regional or national plumbing codes as their model code. One such code is the Uniform Plumbing Code (UPC), as prepared by the International Association of Plumbing and Mechanical Officials, Walnut, California. The UPC, in part, specifies building sewer materials, appurtenances, constuction grade, bedding materials, fittings, and pipe joining. It also includes a moratorium on the introduction of materials or wastes that could damage the sewer (Section 714.1) and wording that "no rain, surface, or subsurface water shall be connected to or discharged" (Section 714.2). However, the UPC does not provide language associated with enforcement of the code nor the assessment of fines and penalities and administrative processes for dealing with non-complying property owners. These issues are the responsibility of the municipality when writing its local sewer ordinance. Many municipalities reserve the right to make corrections to defective private services after giving the property owner adequate notice of the need to repair the service and identifyng the municipality's intention to make corrections at the cost to the owner. These same ordinances provide a process for property owners to appeal the municipality's position or seek permanent or temporary relief from any further enforcement actions.

Enforceability of Correction Orders to Individuals. Voluntary action of property owners to remove illegal inflow sources and repair faulty services is preferred. This helps municipalities reduce capital expenditures in their I/I control strategy and avoid or minimize controversy surrounding expenditure of public funds on private property. Such behavior can be encouraged by furnishing photographs of positive smoke test results to the noncomplying property owner with an explanatory letter of the benefits of correcting these defects or improper (illegal) connections. Additionally, educational materials in the form of sewer bill stuffers or public service announcements can be provided. Nonconfrontational resolution of such matters should always be sought first.

Another avenue for enforcing elimination of illegal inflow sources is through inspections associated with active plumbing permits. Only under this circumstance can a building official coerce a property owner to meet all codes and eliminate all code violations as a term of signing off on the final inspection of new work being effected under the permit. Without an active permit, typically, the authority of a building official is significantly limited in such matters.

The City of Pittsburg, Kansas, initiated an ambitious I/I abatement program in 1989. More than 700 sources of illegal connections on private property were targeted for action. In terms of enforcement, the program far exceeded expectations. Approximately 50% of the affected property owners elected to correct problems with no further intervention by the city. The other 50% pursued reimbursement by the city for the cost of disconnection. Fourteen owners were summoned to appear in municipal court, and only one owner was levied a court fine and issued an order to disconnect. The program was successfully completed in 1992 (Wade et al., 1994).

In implementing a program such as Pittsburg's, one area of immediate concern to a municipality relates to those property owners who may choose to challenge its constitutionality, causing the municipality to incur legal costs to defend its actions. Although a municipality's threat to inspect a home for illegal connections, disconnect illegal inflow sources, or correct defective services while imposing liens on the property to recover costs may be politically unpalatable and the idea of terminating water service unacceptably severe, case law has found such actions not an unconstitutional abuse of power. However, any ordinance writing undertaken to implement new or strengthened language addressing I/I abatement should be carefully guided by an attorney to ensure "due process" (of law)—the course of legal proceedings established by the legal system to protect individual rights and liberties.

Owner/Tenant Responsibilities. An owner/operator of a sanitary sewer system cannot exclusively depend on mandates, rules, codes, or regulations to bring about accountable citizenship and responsible use of its sanitary sewer system. Furthermore, the threat of such enforcement actions as fines and, worse yet, termination of service as exclusive measures to obtain compliance is unreasonable and unrealistic. Additionally, with significant quantities of rental property in all communities around the country, the property owner should not be the sole focus of the municipality's efforts to gain compliance. Tenants must be equally informed of acceptable practices. Ignorance is no excuse for an alleged "innocent" connection of an area drain so a tenant may reclaim a "wet" portion of the landlord's yard for personal use. Yet a tenant should not be responsible for repair costs that are duly the property owner's.

Home ownership includes all of the responsibilities of proper maintenance and protection of the structure and its electrical, mechanical, and plumbing systems—indoors and outdoors. Covenants of the mortgage require this. A public education program should be aimed at property owners and their tenants alike, identifying the responsiblities that must be honored for the privilege of having a collection system in good, reliable working order to handle waste.

Code/Ordinance Requirements and Possible Lack of Uniformity. It is likely that regional model plumbing codes and sewer ordinances used in one part of the country will contradict certain aspects of those used in other parts of the country. Differences in climate (sunbelt vs north tier), topography (flat vs hilly), or predominant soil types (sands vs clays), among other factors, will drive variations in plumbing design standards, specified materials, and methods of testing. The handling of "special" wastes could vary as well. For example, swimming pool overflows and air conditioning condensate discharges may or may not be allowed to be discharged to the sanitary sewer. Additionally, many model codes provide for "administrative" approval, on a case-by-case basis, of homeowner practices that would otherwise be considered illegal connections. Typically, however, sources of wet weather or groundwater-related discharges such as area, basement, and foundation drains and roof gutters will universally not be allowed to discharge to a sanitary sewer. Careful consideration must be given to all features of a model code with respect to local standards of practice and expectations of the local regulators before their adoption.

Health Effects

Diseases Transmitted by Wastewater. A summary of disease-causing biological agents present in wastewater and their associated diseases in humans is given below (U.S. EPA, 1992).

- Bacteria: cholera, salmonellosis (food poisoning), typhoid fever, bacillary dysentary, gastroenteritis (diarrhea and abdominal discomfort);
- Viruses: hepatitis, meningitis, pneumonia, fever, common cold, paralysis, encephalitis, gastroenteritis, diarrhea, respiratory infections;
- Protozoa: gastroenteritis, acute enteritis, giardiasis, dysentery, toxoplasmosis, cryptosporidiosis;
- Helminths: digestive and nutritional disturbances, abdominal pain, vomiting, restlessness, coughing, chest pain, diarrhea, anemia, weight loss, fever, muscle aches, nervousness, insomnia, anorexia, hookworm disease, taeniasis; and
- Bioaerosals: allergic reactions (e.g., asthma) and Legionnaires' disease.

Direct Exposure to Overflows/Basement Backups. Studies have documented incidences of waterborne diseases in humans as a result of contact with contaminated water used for recreation. Such contamination is typically from sewer breaks and overflows or failing on-site wastewater treatment systems (e.g., septic tanks and cesspools) whose discharges reach rivers, streams, creeks, or other bodies of water that are tributary to waters in which full-body contact occurs. In the case of wet weather SSOs, there is significant dilution of wastewater. Dilution reduces the risk of contracting a disease, even though the disease-causing agent may be present. Annually, hundreds of days of use of public recreational water are lost to mandated closures because of possible wastewater contamination (U.S. EPA, 1996).

In contrast, when human contact is made with raw wastewater in a flooded basement, saturated lawn, street gutter, or drainage ditch, the possibility of contracting a waterborne disease is substantially increased in contrast to the risk associated with contaminated recreational water in which substantial dilution of wastewater is likely. Furthermore, adults will typically avoid overflowing sewers and water of suspicious origins, whereas children are more likely to come in direct contact with contaminated water because of an absence of discretion and judgment. This increases the population at risk.

Fish/Shellfish Contamination. Shellfish for human consumption contaminated with wastewater is another pathway for contraction of disease. Local electronic and printed media frequently provide incidences of reported cases of disease and the risk to health in consuming uncooked or improperly handled or prepared fish or shellfish. Reported cases of illnesses from shellfish consumption and as other foods, including meats, poultry, fruits, and vegetables, are greatly increasing. One study suggests that the number of unreported cases could be as high as 20 times reported cases. For the very young and old and those individuals suffering from immunosuppressive diseases or other illnesses, contracting a disease from a wastewater pathogen is a real threat to their lives (U.S. EPA, 1996).

Water Resource Degradation. The potential effects of SSOs are multifaceted. The single effect of greatest concern is that associated with the well being of humans who may come in direct contact with contaminated water or whose exposure is indirect such as through the pathway of an infected foodhandler who has not practiced responsible hygiene. These health concerns are incontestably supported by documented accounts of diseases and illnesses in humans attributable to wastewater-derived, waterborne pathogens.

However, SSOs that discharge to streams, rivers, lakes, bays, estuaries, and oceans also threaten the economic vitality of an area. For example, chronic overflows to bodies of water that support commercial fisheries could result in continuous postings by regulatory agencies, depriving fishermen and workers of related industries of their livelihood. This, in turn, could cause "downstream" economic effects by removing that purchasing power from the local economy. Tourism and real estate values could be negatively affected by real or falsely perceived threats to consumer interests. Industries and manufacturing dependent on a minimum level of water quality for their respective needs may incur added costs to treat the raw water supply, affecting their competitiveness in the marketplace and possibly motivating them to relocate themselves. Furthermore, large employers investigating a particular location for future expansion of their business activities may pass on an area where there are unresolved water quality issues attributable to SSOs that could affect the quality of life for their workforces.

Enforcement

Integral to eliminating or reducing commonly committed mistakes during installation of service laterals and resulting I/I-causing defects is a program that requires homeowners and their contractors to adhere to rules and regulations controlling their installation and minimum material specifications. Typically, adoption by local ordinance of one of several available national or regional model plumbing codes with no further modification will suffice and is a cost-effective way to implement such a program. In addition, these plumbing code requirements must be adequately supported by defendable ordinances to facilitate recourse and enforcement of penalties associated with code violations. Fees, fines, assessments, property liens, and termination of service are examples of penalties that have successfully withstood attack and litigation by grieving property owners in various local, state, and federal court cases.

Key Elements. The fundamental components of a program to ensure that minimum requirements are met in the installation of service laterals include

- Permitting process for any property owner who wishes to connect to the sanitary sewer collection system. No connections should be made until the private system has been inspected and its testing witnessed by the collection system owner's building official.

- Maintained list of licensed plumbers and general contractors approved to perform all original installations, modifications, and repairs to existing service laterals.
- Codified minimum specifications for pipe and construction materials and related appurtenances.
- Regulated list of approved wastewater discharge devices that can be connected to the sewer with stormwater (e.g., roof leaders) or groundwater collection (e.g., foundation drains) or discharge devices (e.g., sump pumps) precluded from such connection.
- Regulated list of discharge devices that cannot be connected to the sewer, including roof leaders, foundation drains, sump pumps, floor drains, and sewer services below the basement floor.

Inspection. The service lateral should be inspected before connection to the sewer main line. A sewer inspection report should be filed with the department of the collection system owner responsible for checking plumbing plans, issuing plumbing permits, and conducting inspections. A typical plumbing inspector checklist/test report is shown in Figure 4. Collection system owners should consider conducting their own lateral service installation inspections instead of having such inspections furnished under contract or by another subdivision of government. This will afford those inspectors a working familiarity with installation methods, service locations, and material specifications, which will facilitate future inspections for private property compliance with I/I control features of the sewer code. The videotape titled "Building Sewer Installation and Inspection" (Water Environment

1. Owner	2. Property Address

3. BUILDING TYPE

□ Single Family Dwelling □Multi-Family Dwelling □ Commercial

4. BUILDING SEWER PIPE □ 4" □ 6" □ ________________ (Other)

□ Extra-Strength V.C.P. ASTM C-200

□ Sch 40 PVC Sewer Pipe (residential only) ASTM D-1785

□ Service Weight C.I. Soil Pipe ASTM A-74-66

□ Sch 80 PVC Sewer Pipe (residential only)

5. JOINING MATERIALS

□ V.C.P Compression Joints ASTM C-425, Type III

□ C.I.P. Neoprene Rubber Gaskets C.I.S.P.I. HSN-68T

□ C.I.P. Leaded & Caulked joints

□ PVC-Solvent Welded Joints
Solvent-ASTM D-2564
Fittings-Sch 40 ASTM D-2466
Sch 80 ASTM D-2467

6. CONSTRUCTION

If adapters are needed, are they of approved type □ Yes □ No

Uniform Bearing □ Yes □ No

Floor Drains Connected □ None □ Permit Issued

Cleanouts □ Adequate □ Inadequate

Trenching Conditions □ Normal □ Special

Sewer Grade □ Adequate □ Inadequate

Change in Sewer Direction □ Adequate □ Inadequate

Obstruction to Flow □ None □ Yes

Leaded Joints □ Proper □ Improper

House Traps □ None □ Properly Installed

Backfilling of Trenches □ Observed □ Not Observed

7. TEN FOOT STATIC WATER HEAD LEAKAGE TEST

□ First Test □ Retest □ $_____ Retest Fee Paid □ Passed □ Failed □ Not Tested

8. REASON FOR FAILURE OR NOTABLE FEATURE:

____

____

9. By reason of the Inspection performed, the Building Sewer located on the above property has been

□ Approved (Only after sewer system is in operation and final connections made)

□ Disapproved (Until building is connected to sewer system)

□ Disapproved (reinspection required)

Date______________________ By_________________________________

Title_________________________________

Figure 4 Building sewer inspection report.

Federation, Alexandria, Virginia) is one source for inspector training.

A multiple inspection format for each service installation may be considered with the following steps:

- Inspect plumber/contractor installation;
- Perform private service connection testing after site grading is completed as a stipulation for issuing certificate of occupancy;
- Television inspect service connections that fail air and water testing;
- Retest service connections and perform building inspection as a requirement of building resale; and
- Provide new building owners with information describing legal and illegal private service connection discharges for future reference.

CASE STUDY 1—VARIOUS COMMUNITIES IN THE SOUTH CENTRAL UNITED STATES

Introduction

Sewer system evaluation surveys of various cities in the south central United States performed in the past few years indicate private sector defects. These defects, which are cost effective to eliminate, produce an average of approximately 20% of total cost-effective inflow. (Cost-effective sources are those for which the cost to repair the defect is less than the cost to transport the inflow through the system and treat it at the plant for a selected planning period.) Table 6 presents total inflow to the system during a rainfall event with a 1-year-60-minute intensity, total cost-effective inflow, and total cost-effective private sector inflow for selected cities. Cost-effectiveness is calculated based on estimated cost for a contractor to repair private sector sources.

Cities are developing programs to eliminate private sector inflow defects that are significant contributors to SSOs, high sewer treatment cost, and plant capacity problems during wet weather. The City of Tulsa, Oklahoma; the City of Dallas, Texas; the City of Fort Worth, Texas; and many other cities have fully met this

Table 6 Total inflow versus private sector inflow.[a]

Study	Total inflow, m³/d	Cost-effective inflow, m³/d	Cost-effective private sector inflow, m³/d	Cost-effective private sector inflow, %
Ft. Worth, Texas, M-272D2	18 652	9 061	1 370	15.1
Ft. Worth, Texas, M-219	13 456	7 805	1 003	12.9
Ft. Worth, Texas, M131A	7 316	7 316	1 525	20.8[b]
Arlington, Texas, basins 24, 25, and 26	7 460	2 021	802	39.7
Irving, Texas, basin 8	6 203	4 368	299	6.9
Highland Village, Texas	833	757	87	11.5
Tulsa, Oklahoma, Coal Creek Basin	80 590	49 711	13 455	27.1
League City, Texas	42 763	25 760	5 098	19.8
Dallas, Texas, Woody Branch	25 488	13 103	3 096	23.6
Irving, Texas, basin 1 and basin 2	9 326	3 940	273	6.9
Johnson County, Kansas, phase 1	51 442	21 926	7 668	35.0
Siloam Springs, Arkansas	22 009	14 632	1 064	7.3
Arlington, Texas, basins 28, 30, and 32	16 620	9 141	2 180	23.8
Lancaster, Texas, basin 8	2 385	1 650	95	5.8
Fayetteville, Arkansas, Illinois River watershed, phase 1	25 529	17 524	867	4.9
Ft. Worth, Texas, M-311	17 093	9 296	969	10.4
Dallas, Texas, Kidd Springs	10 745	5 329	730	13.7[b]
Ft. Worth, Texas, M-269	9 352	4 311	537	12.5
Ft. Worth, Texas, M-216	13 232	4 315	1 397	32.4
Irving, Texas, basins 5, 7, and 12	20 211	11 158	647	5.8
Total	400 705	223 124	43 162	20.0

[a]Inflow is based on a 1-year-60-minute storm event.
[b]A cost-effective analysis was not performed; therefore, the percentage shown is total private sector inflow versus total inflow.

challenge. These cities' programs include educating the public, providing assistance to repair defects, hiring contractors with city funds to repair defects, and enforcing repairs with existing and new ordinances.

Identification of Private Sector Defects

From the 20 studies, it seems that the most prevalent cost-effective private sector defects occur in PBSCs, area drains, and cleanouts. Private building sewer connection defects are defective pipe and pipe joints in service from a sewer main to a building and include residential, commercial, and industrial services. Area drains, for example, should be connected to the storm sewer system or drained overland but are connected to the sanitary sewer system instead. Cleanout defects include missing or broken cleanout caps, broken riser pipes, and cracked pipes.

Effects of Private Sector Defects

The average cost-to-inflow ratio of cost-effective public sector sources in the City of Tulsa's Coal Creek collection system is \$66.58 $m^3 \cdot d$ (\$0.252/gpd). For cost-effective private sector inflow sources, the average cost-to-inflow ratio is \$52.84/$m^3 \cdot d$ (\$0.200/gpd). These values are calculated based on construction cost for a contractor to repair the defects. In lieu of contracting, the city has developed a private sector repair program in which the property owner is responsible for the cost for repairing the defect and which costs the city approximately \$9000 per month to administer. With an average 150 repairs performed in the private sector per month, the cost-to-inflow ratio of the city's portion of the program averages \$5.21/$m^3 \cdot d$ (\$0.020/gpd). Table 7 presents private sector sources in the Coal Creek collection system, amount of estimated inflow generated by these sources, and cost to repair these sources.

The Coal Creek study identified 1169 cost-effective private sector defects. If these sources are eliminated, the city will save approximately \$22 400 per year in transport and treatment costs. Capacity analyses of the Coal Creek system were performed both with and without elimination of private sector defects. All cost-effective public sector defects were eliminated in both analyses. The analysis with private sector defects eliminated reduced capacity improvement cost from approximately \$8.5 million to approximately \$5.6 million, saving the city \$2.9 million in construction costs. The potential cost to the city versus potential savings if private sector defects are eliminated is

Table 7 City of Tulsa cost to repair private sector defects.[a]

Defect description	No. of inflow sources	Total[b] inflow, m^3/d	Cost for contractor repair, \$	City program cost, \$
Cleanouts	664	5 212	117 100	39 840
Building laterals	437	5 019	524 700	26 220
Area drains[c]	48	2 373	57 600	2 880
Sump pumps	11	136	8 500	660
Downspouts	9	712	2 700	540
Total	1 169	13 452	710 600	70 140

[a]Data are from a study for the City of Tulsa in the Coal Creek drainage area.
[b]Based on a 1-year-60-minute storm intensity.
[c]Includes patio drains, driveway drains, etc.

shown in Table 8. These savings represent private sector defect repair in a small portion of the city. As other areas of the city are studied and private sector defects are eliminated, savings will multiply.

Of five studies conducted for the Dallas Water Utilities, which includes approximately 754 km (470 mile) of sewer line, a total of 1257 cleanouts were found to be defective. These cleanouts contributed approximately 5988 m^3/d (1.582 mgd) of 1-year-60-minute inflow. These studies also identified 1437 PBSCs and 26 area drains contributing approximately 9039 m^3/d (2.388 mgd) and 984 m^3/d (0.260 mgd) of 1-year-60-minute inflow, respectively. Using an average cost for collection and treatment, repair of these defects will save an estimated \$19 090 per year in O&M costs and an estimated \$4.23 million in capital cost. The cost to Dallas to repair these defects was approximately \$164 000.

Benefits

There are several benefits to be gained from the repair of private sector defects. As defects are repaired, inflow is eliminated, along with the cost of transporting it through the system and treating it at the plant. Repair of these defects can reduce the number of SSOs that occur during wet weather. Repairs made in the private sector can also reduce inflow enough to eliminate the need for certain sewer system capacity improvements. Many enforcement agencies consider direct connections such as area drains and downspout connections a violation of code/ordinance. Elimination of these defects can ensure the sanitary sewer system is in compliance with applicable regulations.

Private Sector Inflow Defect Elimination Programs

City of Tulsa, Oklahoma. The City of Tulsa, located in northeast Oklahoma, with a population of approximately 370 000, began construction of its sanitary sewer system in the early 1900s. In 1990, the city was issued an administrative order from the U.S. EPA and a consent order from the Oklahoma Department of Environmental Quality for wet weather SSOs in several areas of the city.

In October 1990, the city began an aggressive program to eliminate I/I entering its sanitary sewer system. Sewer system evaluation surveys were begun in several areas of the city in the spring of 1993. As the SSESs progressed, the city realized that private sector defects had significant effects on the sewer system. In August

Table 8 Private sector defects—cost versus savings.

Parameter	Cost	Saving
Defect elimination	$70 140.00[a]	
Capacity improvement[b]		$2 900 000.00
Treatment and O&M cost		$ 22 400.00[c]

[a]Actual cost expended by city program.
[b]Based on 5-year-60-minute storm intensity. Tulsa's design storm standard is a 5-year event. However, individual defects are typically analyzed on a 1- or 2-year event basis. Therefore, this table was adjusted to reflect that standard based on 1-year rainfall intensity of 38 mm/h (1.5 in./h) and a 5-year rainfall intensity of 59.7 mm/h (2.35 in./h).
[c]Annual savings.

1994, the city began revising its private sector defect enforcement program to better meet the needs of the I/I reduction program. Currently, the program is a joint effort of the Public Works Engineering Services and Building Inspections Departments. Public Works Engineering Services administers the SSESs and collects the defect information, manages the information with a computer-assisted defect inventory, and routes defect information to Building Inspections. When the defect is repaired, the defect is removed from the inventory.

The city has assigned one full-time building inspector and one part-time building inspector, each with a plumbing license, to the program. Building inspectors are used in Tulsa because they are authorized to write citations. One city vehicle is used for the program. Other miscellaneous expenses such as printing, postage, and supplies are covered by the program.

Forms with sketches showing approximate locations of the private sector inflow defects, created during the SSES projects, are provided to the inspectors by the Public Works Engineering Services Department. Photographs of the defects are taken during the SSES and also given to the city to aid verification and location. The inspector visits each defect site. If the owner-occupant is unavailable at a defect location, the inspector leaves an official notice indicating the location of the defect, a business card, and a letter describing the I/I elimination program. If the defect is difficult to locate, the inspector spray-paints the defect location. The notice gives the owner-occupant 7 days to take action to repair the defect. If a permit is taken out to repair the service, the inspector notes the date. If a cleanout cap is required, the inspector revisits the site after 7 days. If the defect requires a permit and no permit has been applied for, the inspector also revisits the location. If on the second visit the defect has not been repaired, the inspector leaves another notice giving the owner-occupant 24 hours to respond. If after approximately 48 hours have passed (after the second visit) no action has been taken to correct the defect, the city sends a registered letter stating that the water meter will be pulled if action is not taken immediately. Ten meters have been pulled in the City of Tulsa since the program began.

The inspector spends approximately 30 minutes at each location with a defect and averages 200 visits per month. The city estimates that approximately 80% of defects are corrected after the first notification. Of the sites visited, approximately 25% have an occupant at the site. At the sites where notices are left, approximately 90% of the owner-occupants call the inspector, and 10% of those want to meet with the inspector.

City of Dallas, Texas. The City of Dallas, Texas (the eighth largest city in the United States), began its program to eliminate private sector I/I defects in 1987. The program is administered by the Dallas Water Utilities. A crew of two and one vehicle are used for the program.

The City of Dallas meets with every occupant-owner with a private sector defect. If no one is present on the first visit, the city leaves a notice to call and set up an appointment. The city uses completed smoke testing forms and pictures taken during smoke testing to show defects to property owners. Crews visit each site as many as three times to ensure the defect has been corrected. The city has an approximate 95% success rate. Pursuit of approximately 5% of the defects that have been found to be minor and expensive to fix are deferred. The City of Dallas also provides cleanout caps and plugs sewer services to abandoned buildings at the main.

In areas where SSOs are apparent, response time from property owners is much shorter. The city does not observe a difference in its success rate in low-income and high-income neighborhoods. The city has used its ordinances twice, threatening to pull two meters in the 9 years since the program began. The defects were repaired and the meters did not have to be pulled. According to city staff, the city's program is capable of eliminating approximately 1800 private sector defects per year, which would eliminate approximately 9840 m³/d (2.6 mgd) of inflow (1-year-60-minute storm).

City of Fort Worth, Texas. The City of Fort Worth began its I/I elimination program in 1988, performing comprehensive SSESs, designing corresponding public sector rehabilitation, and completing recommended repairs. In 1993, the city was placed under administrative order by U.S. EPA to eliminate reported SSOs, although its program was already successfully eliminating SSOs in the areas studied. Table 9 shows the cost to repair private sector defects in several basins in the City of Fort Worth.

The three construction companies contracted by the City of Fort Worth for each of the projects were provided with a list of addresses where the defects were located and a description and picture of each defect with specifications on repair. The contractors contacted each homeowner for permission to work on the property. When written permission was received, the contractor repaired the defect. Each of the three contractors successfully completed approximately 95% of the repairs. The 5% that were not repaired included residents who could not be reached and those who refused access to cleanouts or yards. The private sector repairs in drainage basins M-269, M-311, and M-272D2 should eliminate 3040 m³/d (0.76 mgd) of

Table 9 City of Forth Worth average cost to repair private sector defects.

Description	Basin M-269	Basin M-311	Basin M-272D2	Average
Cleanout cap	$120.00	$ 15.00	$ 2.00	$ 46.57
Cleanout housing and riser	$150.00	$230.00	$ 85.00	$155.00
Service lateral repair	[a]	[a]	$200.00	$200.00

[a] Not included in contract.

inflow (1-year-60-minute). Postrehabilitation flow monitoring was conducted in the fall of 1996 or spring of 1997 to verify the effectiveness of eliminating the defects.

CASE STUDY 2—LOWER PAXTON TOWNSHIP, HARRISBURG, PENNSYLVANIA

Introduction

The Lower Paxton Township Authority in Harrisburg, Pennsylvania, was faced with a serious I/I problem caused by, among other things, many years of development without a rigid inspection program for installation of PBSCs. (The authority constructed the sewer mains and the service laterals extending from the mains to the edge of right-of-way, where the PBSCs were made.) With a system of more than 340 km (210 mile) of sewer mainline and 14 000 customers, the 42 000-population township was experiencing peak wastewater flow of more than 10 times average daily flow. A settlement agreement and consent order developed among the Pennsylvania Department of Environmental Protection, housing developers, and a downstream municipality required the entire system to be rehabilitated within a 6-year period, with penalties for failure to achieve specified measurable criteria.

Methodology

The sewer system contains approximately 340 000 m (1 110 000 ft) of sewer and four pumping stations. It is divided into four drainage basins: Beaver Creek, Paxton Creek, Spring Creek, and Asylum Run. Significant occurrences of wastewater overflow at manholes and excessive flow entering downstream municipalities were documented in 1989 for Spring Creek and in 1993 for the Beaver and Paxton Creek basins. Peak storm event flow of 76 000 to 91 000 m^3/d (20 to 24 mgd) in the Beaver and Paxton Creek basins was 3 to 4 times the capacity of downstream interceptors, whereas the Spring Creek peak flow was approximately 1/2 times downstream capacity.

The authority's rehabilitation program includes comprehensive flow metering, wet weather physical survey and internal televising, meetings with customers to inspect for floor drains and sump pumps, rehabilitation of manholes and sewers, repairs to PBSCs, removal of area drains and sump pumps, and installation of clear-water systems in neighborhoods to properly convey sump pump discharges. Program administration includes contract management, data management and analysis, selection of appropriate sources for removal, design engineering, and reporting. Significant activities are tracked through an information management system (IMS) and reported through a graphical information system (GIS). The IMS is a database system that township staff can access via modem, and the GIS was developed from base mapping provided by the Pennsylvania Department of Transportation and existing sewer system drawings. Locations of many key elements of the sewer system were field verified using global positioning system surveys. The IMS and GIS were integrated to generate progress maps and sort data for decisions on field investigations and rehabilitation.

During 1994 and 1995, detailed physical surveys, internal television inspections, and private source inspections were performed extensively in the four demonstration subbasins. Rehabilitation of the public sewers was performed in these subbasins to determine the effect on flow reduction and the degree to which groundwater migration would increase source flow from private properties. The private source inspections included building inspections of all properties within these subbasins, water saturation tests along the alignment of PBSCs, replacement of the PBSC to the public lateral at the curb line, and internal television inspection of PBSCs during rainfall and other high groundwater events. Flowmeters were installed at key locations throughout the system to monitor basin flow and identify wet weather events suitable for investigation.

Flow Analyses

Results of investigations in the demonstration areas indicated that private sources of I/I are significant. Sources, measured before any rehabilitation, are distributed approximately as shown in Table 10. Approximately 40% of the I/I were initially observed in the public sewers, and 60% were identified as coming from PBSCs (including contribution from illegally connected sump pumps and area drains). After rehabilitating the main sewers, approximately 90% of remaining I/I flow have been observed to be coming from the private sector.

Target Removal Goals

Given a target removal goal, the efficiency of the source removal can be calculated and I/I can be quantified after deducting theoretical use. Of the 40% of I/I located in the public sewers (27 800 m^3/d [7.5 mgd]), it was found that approximately one-third of this flow is effectively removed from the system. The other two-thirds of the extraneous flow that was previously associated with the public sector migrate through the groundwater to PBSCs. This increases private sources from the initial 60%, or 41 700 m^3/d (11.3 mgd), to a total of approximately 57 000 m^3/d (15.5 mgd). To achieve a goal of 18 700 m^3/d (4.9 mgd) (the capacity of the interceptor sewer leaving the

Table 10 Distribution of I/I sources.

Source	I/I flow contributed during a peak event, %
Public sources	
Manholes correctable through grout	25
Sewers correctable through grout	14
Sewers requiring external repair	<1
Private sources	
Sump pumps to be removed	19
Building sewer repairs	42

township), only 14 200 m^3/d (3.7 mgd) of I/I may remain with the base usage of 4500 m^3/d (1.2 mgd). In this case, approximately 80% of private sources must be removed to achieve the acceptable quantity of I/I without exceeding system capacity.

The conditions of the consent order negotiated with the Pennsylvania Department of Environmental Protection required specific measurable criteria to be met before a limited ban on new connections would be lifted. The measurable criteria included reducing average wastewater flow per equivalent dwelling unit (EDU) to approximately 950 L/d per EDU (250 gpd/EDU), reducing the peak-to-dry-weather flow ratio to 2.5 in the Beaver Creek basin and eliminating all of the SSOs. As a result, rehabilitation of all of the public I/I sources and removal of 80% of the private sources were required.

Rehabilitation Costs

To assign relative cost effectiveness to each area of rehabilitation, the cost of each type of repair was evaluated versus quantity of flow anticipated to be removed. Alternative techniques included methods of eliminating overloaded sewers by providing equalization and storage for SSOs and/or construction of additional wastewater treatment capacity. Rehabilitation of private sources was priced fully regardless of whether the property owner or public agency would be paying for repairs. Results are shown in Table 11 on a present worth basis.

Cost estimates for repairs to PBSCs were prepared based on total replacement. This included a cost of \$130/m (\$40/lin ft), which would exclude surface restoration. An estimated cost of \$2000 for installation of a permissible sump pump, if none exists and a property has basement service, was also included. The mean cost for all leaking building sewers is approximately \$3700. For PBSCs with flow greater than 3.8 m^3/d (1000 gpd), the mean cost is approximately \$3600.

Cost Effectiveness

The decision to select certain sources of identified I/I for repair must be based on specific criteria to ensure

Table 11 Estimated present worth costs associated with available rehabilitation techniques.

Control option	Unit present worth cost	
	1996 $/L	1996 $/gal
Aboveground overflow tank, 5680 m^3 (1.5 Mg)	$ 0.25	$ 0.98
Belowground overflow tank, 380 to 6050 m^3 (0.1 to 1.6 Mg)	$0.35 to 1.27	$1.32 to 4.80
Purchase additional capacity at existing WWTP	$ 1.63	$ 6.16
Construct new WWTP, 380 to 3800 m^3 (0.1 to 1.0 Mg)	$1.00 to 4.97	$3.81 to 18.80
Remove I/I sources		
Public facilities		
Manhole grouting	$0.01 to 0.08	$0.04 to 0.30
Main line grouting (by LPTA)	$0.01 to 0.10	$0.05 to 0.36
Main line grouting (by contractor)	$0.06 to 0.39	$0.21 to 1.48
Main line replacement	$0.04 to 0.59	$0.16 to 2.22
Private facilities		
Sump pump/drain removal	$0.01 to 0.07	$0.20 to 3.72
Building sewer replacement	$0.05 to 0.98	$0.20 to 0.27

that only the most cost-effective repairs are made. These criteria include

- Unit cost of removal or source. Unit costs of removal were described earlier and are summarized in Table 11.
- Efficiency of removal or source. Because of the migration of groundwater, the efficiency of public sewer rehabilitation in a residential neighborhood may be reduced by approximately one-third.
- Best order of source removal. The normal practice of first addressing public sewer repair should be maintained. This practice demonstrates to the public that private lines are not

being unduly targeted before proper maintenance is performed on public facilities. Also, public sewers are typically buried deeper and, if not addressed first, will be more severely affected by the elevated groundwater caused by other rehabilitation activities.

- Cost comparison with other options. As more I/I sources are removed, other sources must be addressed that will have greater unit costs to repair. As unit costs increase, a decision can be made to consider other options. Rehabilitation method costs, specifically mainline repairs-replacement and PBSC replacement (only those individual sources with the highest cost per gallon removed), should be compared with the lowest potential cost of a storage-equalization facility.

The goal of following these criteria is to accomplish necessary flow reduction at the most favorable cost. This methodology requires the assessment of cost effectiveness and establishing of an initial benchmark.

Flow Thresholds for Rehabilitation

To simplify data analysis, average values for each flow level were determined and graphed. This was used to determine a flow threshold below which other solutions may be more cost effective. For instance, storage-equalization facilities sized from 1900 to 3800 m^3 (500 000 to 1 000 000 gal) have an estimated present worth of \$0.40 to \$0.53/L (\$1.50 to \$2.00/gal). On this basis, an initial threshold of approximately 7.6 m^3/d (2000 gpd) may be appropriate. Based on data from the first subbasins, such a threshold would affect approximately 45% of the customers with leaking PBSCs and target 88% of I/I. In 1996, the authority adopted a threshold flow of 5.7 m^3/d (1500 gpd) as the initial target for PBSC rehabilitation orders.

Funding of Private Source Rehabilitation

An extensive review of private source repairs in Pennsylvania and other states was conducted through a literature search and other means. Programs varied from those in which full responsibility rested with the private property owners to those in which full funding was provided by the municipality. Personal communications with municipal agencies conducting private source repairs in Pennsylvania noted a problem with the transition connection from the typically 10-cm (4-in.) PBSC to the 15-cm (6-in.) public service lateral at the street.

Lower Paxton Township initially elected a program with 50% reimbursement for repairs performed by property owners; however, the schedule was hampered by slow or nonresponsive property owners and plumbers. So the township created the current program, which involves a contractor directly employed by the township authority with the full cost of repairs borne by the authority. Before PBSC building sewer rehabilitation, the authority requires a repair and rehabilitation agreement, which covers access to the property, indemnifications, and maintenance of the line by the property owner. The program modification has increased participation from 22 to 95%. The program was also enhanced by additional provisions for testing each PBSC after performing the minor repair of replacing the transition connection with a test tee. Based on pilot testing, it was anticipated that 35 to 40% of PBSCs would pass an air test after the transition connection was replaced. This program enhancement reduced average cost of each repair from approximately $3600 to $2000. Results of the program to date are that 55% of the PBSCs required only replacement of the transition. Note that the program already required removal or rerouting of any area drain or sump pump illegally connected to the sewer system.

Results

Effects of I/I on subbasins were evaluated through development of hydrographs of actual flow during peak rain events versus average base flow to graphically represent the response of system flow relative to rainfall and groundwater level. Only comparable rainfall events were analyzed (approximately 7 to 10 cm [3 to 4 in.]). As of December 1996, with only part of the private source program completed, four subbasins showed significant flow reductions. For these subbasins (BC-1, BC-1/2, BC-3, and BC-6), prerehabilitated and current flows were compared, producing the following approximate percent reductions in I/I: 90% (from 15 000 to 1 500 m^3/cm [1.0 to 0.1 mil. gal/in.]), 56% (from 29 000 to 13 000 m^3/cm [1.95 to 0.9 mil. gal/in.]), and 81% (from 50 000 to 9500 m^3/cm [3.36 to 0.64 mil. gal/in.]), respectively. In addition, periods when flow exceeded pumping station capacity in one subbasin were reduced 94% from 384 hours (16 days) to 22 hours (0.9 days) during the course of the program. The significance of the flow reductions made in these areas is highlighted by the fact that the flows measured during the postrehabilitation period (1996) occurred during the year with the most precipitation in the past decade for central Pennsylvania.

Conclusions

The I/I program in Lower Paxton Township has demonstrated the cost effectiveness of rehabilitating private sources of I/I. The program has not only regained capacity lost because of extraneous flows but has also, as a result of the stringent criteria set for the program, "created" capacity not previously planned for in the system design. Key concepts evident from this case study include the following:

- As reported by other programs, private sources in Lower Paxton were calculated initially to make up 60% of system I/I.
- After rehabilitation of public sources of I/I, groundwater migration causes private sources to represent up to 90% of the original I/I in the sewer system.
- Cost-effectiveness techniques demonstrate the value of targeting private source rehabilitation.
- An integrated GIS–IMS is a powerful and cost-effective tool for optimizing

sewer system investigation and rehabilitation programs.

- Private source repairs can be less costly than construction of additional conveyance and treatment capacity.
- Program economics are optimized by quantifying PBSC flow and establishing thresholds for ordering rehabilitation.
- PBSC rehabilitation programs should include transition connection replacement with subsequent air testing of the sewer before ordering total line replacement.

A carefully regulated and managed program for private source rehabilitation with full public funding can meet aggressive goals and schedules and can be accomplished cost effectively.

CASE STUDY 3—JOHNSON COUNTY UNIFIED WASTEWATER DISTRICTS

Introduction

Johnson County Wastewater (JCW) serves more than 250 000 people within a rapidly growing area of Johnson County, Kansas. Currently, JCW operates nine wastewater treatment facilities throughout the county, treating an average of approximately 132 500 m^3/d (35 mgd) of wastewater. It also cleans and maintains approximately 2440 km (1515 mile) of sanitary sewer and more than 32 000 manholes. Since 1945, several main, sub, and joint sewer districts were created, with the oldest and largest districts being the Mission Township Main Sewer District no. 1 (Mission Main), Indian Creek Main Sewer District no. 1 (Indian Creek), and Shawnee and Mission Township Turkey Creek Main Sewer District no. 1 (Turkey Creek). These districts contain more than 51% of the total linear feet of sewers and approximately 18 000 manholes.

The sanitary sewer system in Johnson County began experiencing both excessive groundwater infiltration and stormwater inflow. Excessive I/I was entering the sewers during storm events, which often led to surcharged manholes, raw wastewater bypassed to receiving streams, and wastewater backups in basements. Not only did the I/I cause considerable financial loss each year but also constituted a potential health hazard. During heavy rainfall events, the districts would receive numerous phone calls from residents concerning basement flooding and wastewater backups. In 1983, the Board of County Commissioners of Johnson County decided to embark on an extensive public and private I/I reduction program.

Sanitary Sewer Evaluation Study

Each study conducted involved several areas of investigation and analysis. The activities surveyed typically included system flow determination and identification and quantification of I/I sources. Specific activities investigated included flow and rainfall monitoring, manhole and visual pipe inspections, building inspections, smoke testing of sewer lines, dye water testing of private sources, and television inspections. During each study, intensive surveys were performed to evaluate physical conditions of manholes and sewer lines, and building inspections were conducted by the consultant to record data such as building type, age, and use; flooding history (source, occurrence, and frequency); foundation type and dimensions; type of source and route of discharge; downspouts (number and route of discharge); suspect area drains; and lot drainage adequacy. Table 12 summarizes results of the building inspections. Smoke testing was also done to quickly detect inflow sources such as downspouts, area drains, foundation drains, abandoned PBSCs, faulty connections, direct connections, and storm sewer cross connections (see Table 13). Dye water testing was also performed on certain suspect sources to confirm actual PBSCs.

Based on the detailed analysis for each study, a recommended plan was developed for I/I elimination, new sewer lines, and related pumping station and treatment plant upgrades. Reports indicated that it was cost effective to remove 4468 private sources from the Mission Main District, 9623 sources from Indian Creek, and 5711 sources from Turkey Creek. Inflow from these private sources represented an average 69% of total inflow in these three districts, and 45 to 50% of this inflow was cost effective to remove.

Private Infiltration and Inflow Removal Program

Johnson County Wastewater elected to create a private I/I removal program, which began with the development of a resolution to establish the program. This resolution was approved in 1985 and amended in 1992 and contained the following primary provisions:

- That excess I/I overloads the existing sanitary sewer system and, therefore, is a hazard to the health and welfare of the community;
- That any property owner, properly notified, is required to disconnect an I/I source as described in the resolution;
- Definitions of infractions and penalties of the resolution and procedures for gaining access to properties for testing and inspections;
- That on verifying that the I/I source has been disconnected from the sanitary sewer, the property owner

Table 12 Building inspection results.

District	Total buildings	Sanitary sewer backups	Suspect inflow sources	Confirmed inflow sources
Mission Main	18 059	1 472	13 801	637
Indian Creek	16 655	1 224	9 182	5 038
Turkey Creek	16 668	719	8 896	2 298

Table 13 Smoke testing data.

Type of source	Mission Main	Indian Creek	Turkey Creek
Confirmed			
Building service line	72	230	57
Cleanout	76	425	181
Transition joint	11	70	24
Area drains	80	329	127
Foundation drains	64	237	240
Downspouts	6	53	—
Suspect			
Area drains	1822	664	2175
Potential cross connection with storm	41	51	32
Total	2172	2059	2836

may apply for reimbursement of all directly related costs, up to defined maximum amounts; and

- A process for review and appeal of disconnection orders.

The primary goal of the project was to create a program that would disconnect the source identified by inspection in a prompt and orderly manner. This program would also use local plumbers and contractors for the disconnections and assist property owners with the process whenever possible. Commencing in 1986, the program's original staff consisted of two inspectors and a planning coordinator. The inspectors' duties included dye water testing of unconfirmed suspect sources, inspection of vacant or uninspected buildings, and inspection and verification of disconnected sources. The program eventually expanded to nine staff members, primarily to improve the disconnection rate because further inspections found more sources than were originally projected.

At project startup, several objectives were established to ensure that the project was manageable

- Construction of a dye water trailer for follow-up testing,
- Establishment of procedures for testing and inspecting sources,
- Tracking of the testing and disconnection process, and
- Establishment of minimum standards for source disconnection.

The size of the project—that is, the number of sources to be disconnected, tested, and inspected—required that an alternative to paper files be devised to track the process from start to finish. A computer tracking system based on a relational database was designed in-house to handle every aspect of the process, including

- An easy-to-use data entry screen and record storage system that allowed easy record access to answer inquiries and update records,
- Computer-generated disconnect notices and periodic reminder letters to property owners,
- Computer-generated requests for access and periodic reminder notices to perform building inspections,
- Computer-generated lists of suspect sources to be dye water tested,
- Programs to handle ASCII file transfers of property ownership data from the county's mainframe real estate records, and
- Complete monthly and annual reporting of all activities related to the program, for example, monthly expenditures, number of sources disconnected, a list of participating contractors, an executive summary of activities, etc.

Before property owners were notified to disconnect I/I sources, standard construction methods and minimum standards for construction materials were established.

This was done to ensure that all disconnections would comply with local plumbing, electrical, and building codes and to assure property owners that a quality job would be performed.

Source disconnections performed during this project involved 10 separate city governments, each having code enforcement authority within their borders. Many of the disconnections would require modifications to existing interior plumbing and electrical systems. Meetings were held with city representatives before project startup to determine whether normal permit and inspection procedures would be followed. These discussions led to the waiving of individual city permits, permit fees, and inspections to reliance on district inspectors to enforce code compliance.

To reduce the time it took to disconnect a source, staff devised a method for establishing an informal fixed-price contract with local contractors. Specifications and set costs were established for disconnection of foundation drain sump pits, which made the process much easier for all involved parties. Owners were given the option of calling the districts and requesting that the work be assigned to a preapproved contractor. If they did not wish to follow this method, a list of contractors was mailed to them so that they could proceed with a two-bid process. With this process, time spent by contractors became more productive. In most cases, they only needed to visit the job site to establish a bill of materials for their work crews. Also, inspection of completed work became simpler because standard materials and processes were used.

The effect of this process standardization was remarkable. Public response was positive; homeowners were more confident in the process; and contractors could make efficient use of their resources by scheduling work in clusters within neighborhood areas. Removal rates during October 1990 (the first month that the system was used) increased by approximately three. This standardization allowed rates of more than 4000 per year.

Another innovation was JCW's telephone follow-up system. Early in the program, many owners would not respond to the first notice until more than 60 days had passed and a reminder letter was sent. To reduce the number of reminder letters sent to a minimum, certified return receipt cards were included with the notice. As these were returned, inspectors would call the owners to acknowledge receipt and inquire whether they had questions regarding how to proceed.

Less than 0.1% of homeowners initially refused to perform the work or allow building inspections. Enforcement action was taken as required and most complied. Only nine properties remain uninspected at this time. The updated resolution allowed JCW to transfer the responsibility for enforcement to the county code courts, making final enforcement easier.

Results

The total cost for the entire I/I removal project was $46.8 million, $11.2 million of which was spent on the private I/I program to disconnect sources. Also, JCW received almost $12 million in grant funds and almost $18 million in low-interest state loans. However, private I/I expenditures were not grant or loan eligible and were paid via general obligation bonds. Postrehabilitation flow monitoring was conducted during all stages of the program to verify the amount of I/I removed. Actual I/I reduction for a 10-year storm event for all district phases ranged from 41.7 to 71.1%, which was greater than projected values.

To date, there have been 15 632 sources removed, removing 2.29×10^5 m^{-3}/d (60.5 mgd) of flow from the sanitary sewer system as measured during a 1-year storm event. Eleven homes remain to either be inspected (refused entry) or have a source be disconnected and are being addressed through the legal process.

Current and Future Plans

The JCW has considered expanding the private I/I program to other districts that have begun to show excessive flows. Reinspection of disconnected sources is currently underway. Small SSESs are being conducted in-house, and a flood abatement response team has been formed to respond to and analyze areas that experience collection systems backups. Hydraulic modeling software has been purchased to improve evaluation of the JCW system. Other methods to cost effectively remove remaining sources of I/I from private property (such as indirect foundation drains and service lines) are being reviewed. Flow monitoring, smoke testing, and dye water testing are also being performed currently.

CASE STUDY 4—ROCK RIVER WATER RECLAMATION DISTRICT, ROCKFORD, ILLINOIS

Introduction

The Rock River Water Reclamation District (District) serves the cities of Rockford, Loves Park, and Cherry Valley and the villages of Machesney Park and Roscoe, with certain unincorporated areas, in northern Illinois. The population served by the District is greater than 250 000. The City of Rockford has the oldest sewers, with many sections that are approximately 100 years old. As in many communities with such old sewers, there are areas that experience basement flooding during wet weather. The District has undertaken a systematic program to analyze each problem area on the basis that each area has unique conditions and that the solution to each problem will be site specific. The program attempts to prioritize study areas based on known problem areas or number and frequency of customer complaints. As funding becomes available, each area is studied. System rehabilitation within rights of way and easements is primarily performed by District agents, based on results of the studies.

The purpose of this SSES was to determine the extent of excessive I/I in the basins studied and recommend a plan for mitigating I/I in each basin, including a cost-effectiveness analysis to justify recommended work.

Phase I

General. The District directed a survey of three problematic basins within its service area that were experiencing sanitary sewer backups during periods of heavy rainfall. These backups were substantiated by the building inspection surveys for basins 50 and 121 in Rockford, Illinois, and for basin 157 in Cherry Valley, Illinois, performed in late 1994. (The basins were designated for study as 50 for the Dawson Avenue area, 121 for the Pepper Drive area, and 157 for the Cherry Valley area.) The SSES scope of work was organized in two areas: (1) field investigation and (2) project management, data analysis, and reporting. Specifically, these general work areas are subdivided into the following work tasks:

Field investigation

- Review District plans, specifications, and records.
- Install and maintain six March-McBirney Model 260 (Fredrick, Maryland) meters for at least 8 weeks in the three basin areas.
- Perform physical manhole inspections of 365 manhole structures in the basin areas.
- Provide smoke testing of all (approximately 77 000 lin ft) sanitary sewers to identify I/I sources.
- Perform dye water flooding of storm systems adjacent to sanitary sewers.
- Perform voluntary building inspections of approximately 1300 buildings to identify I/I sources on private property in the basins.
- Properly notify property owners and the public when smoke testing, dye water flooding, and building inspections will be performed.

Project management, data analysis, and reporting

- Provide computer-generated hydrographs and numeric flow data for each flow meter. Provide baseline flow data for each basin.
- Identify I/I sources using the District's internal television inspection data-disks.
- Identify all I/I sources in terms of location, condition, flow rate, rehabilitation method, and cost. List all defects in order of cost effectiveness. Provide ratios of rehabilitation costs versus treatment-transport cost to calculate relative cost effectiveness and provide a basis for rehabilitation recommendations. Provide appropriate color photographs of I/I source to support data.
- Provide a brief summary of findings for each phase and basin area of the survey.

Findings. This SSES found that sanitary sewers were subject to excessive stormwater inflow during wet weather events in basins 50, 121, and 157. It was estimated that 5300 L/min (1400 gpm) inflow (based on 5-year storm frequency) occurs in addition to typical diurnal sanitary flow, causing sewer backups and sanitary sewer surcharging in these basins. Sources of I/I were found in the public sector, that is, from within the street rights of way or easements and sewer defects associated with manholes and sewer lines. Sources were also found in the private sector (on private property), primarily from improper or illegal connections of building foundation drains and storm sump pumps.

A total of 515 defects or improper connections found during the study were identified as sources that would account for approximately one-half (2450 L/min [674 gpm] of the 5300 L/min [1400 gpm]) of the inflow. The primary sources of I/I were foundation drains, sump pumps, manhole lids and frames subject to surface runoff, leaking manhole walls, and roof leaders directly connected to the sanitary sewer system.

Approximately 56% of the primary structures (736 of 1308) were entered and inspected during the building inspection program. Of these,11% (138 buildings), or one of every six residences contacted, refused entry or inspection of the premises. Also, 33% of the buildings were not inspected because residents were not home at the time of the initial survey and could not be successfully contacted during subsequent visits, by mail, or by telephone. Note that 75% of I/I identified and recommended for removal emanated from improper–illegal connections found during the building inspection program. A significant amount of inflow is yet to be located and is likely located on private property that was not inspected during the study.

Other key findings of note are as follows:

- Basin 50
  - The residences in basin 50 were constructed circa 1913 to 1930. The most common type of illegal connections found during building inspections were footing drains tied to a basement floor drain without a sump pit.
- Basin 121
  - The residences in basin 121 were constructed circa 1960 to 1975. The most common type of illegal connections found were sump pumps with footing drains improperly connected to the sanitary sewer.
- Basin 157
  - During the field investigation, it was found that the Elgin Street lift station was operated with the sanitary sewers surcharged during normal operations. Wet well settings

have been adjusted to correct this situation.

Tables 14, 15, and 16 provide summaries of the defects identified and excess flows imposed on each basin.

Conclusions and Recommendations. During the SSES, the field investigation located specific sources contributing an estimated 1900 L/min (510 gpm) of I/I that could be cost effectively removed, 75% of which emanates from private property. Primary sources of I/I are foundation drains, sump pumps, manhole lids and frames subject to surface runoff, leaking manhole walls, and roof leaders directly connected to the sanitary sewer system. Removal of the identified, cost-effective repairs is sufficient to significantly mitigate sanitary sewer surcharging in the study areas. Also, a significant amount of I/I could yet be located, which is likely located on the private properties unavailable for inspection during this study.

Based on the information presented in this report and the field work completed to date, the following actions were recommended:

- The District should remove I/I sources identified within the public right of way that have been shown to be cost effective.
- The District should rehabilitate any structural defects in the sewer lines and manholes, regardless of whether they are found to be cost effective, to maintain the hydraulic capacity and

Table 14 Building inspection survey.

Basin	Number entered	Number refused	Number not home	Reported sewer problems	Improper connection	Total in basin
50	271	55	156	199	26	482
121	277	69	176	160	35	522
157	188	14	102	84	7	304
Total	736	138	434	443	68	1308

Table 15 Identified I/I summary (number of defective sites in parentheses).

District	Manhole lid/frame/ ring, gpm	Pipe joint/line service, gpm	Improper sump/footings/ downspouts, gpm	Total flow, gpm
50	80.1 (84)	63.4 (61)	167.6 (26)	311.1 (171)
121	26.5 (99)	51.7 (92)	147.8 (35)	226.0 (226)
157	49.3 (55)	49.3 (56)	38.0 (7)	136.6 (118)
Total	239.9 (238)	164.4 (209)	421.4 (68)	673.7 (515)

[a]gpm $\times$ 6.308 $\times$ 10^{-2} = 45.

integrity of the sanitary sewer system within the basins.

- The District should report to the City of Rockford and to the village of Cherry Valley, according to the District's intergovernmental agreement, apparent violations of the plumbing code and state regulations for sources of I/I identified on private property. The District should work with the municipalities to see that the improper connections are removed.
- The District should continue its investigation and survey of buildings that were unavailable during this study to complete the survey and further identify I/I sources on private property.

Phase II

General. District sanitary sewer basins 18, 105, 106, 165, and the southern one-half of 107 were included in this SSES. In May 1996, 8 days of rain occurred (May 19 through May 28), culminating in a 10-year rainfall event on May 28. During and following this storm, there were 67 reports of sewer backups and flooded basements in the basins studied, particularly basins 105, 106, and 165 and—to a lesser extent—basins 18 and 107.

Table 16 Flow analysis.

	Baseline flow[a]		Infiltration		Inflow[b]			Total flow			
Subdistrict	Daily, Mg	Annual, Mg	Daily, Mg	Annual, Mg	5-Year/ 1-Hour, mgd	5-Year/ 24-Hour, Mg	Annual, Mg	Peak Hour, gpm	Peak day, mgd	Annual, Mg	Transport capacity, gpm
050–001	0.171	62.42	0.072	9.65	0.756	0.43	3.93	644	0.601	66.35	520 (10 in.)
050–036	0.074	27.01	0.029	3.89	0.488	0.188	1.20	390	0.262	28.21	345 (8 in.)
121–037	0.105	38.33	0.046	6.17	0.518	0.294	1.59	432	0.398	39.38	345 (8 in.)
121–046	0.074	27.01	0.036	4.82	0.329	0.212	1.14	275	0.286	28.46	345 (8 in.)
157–04	0.136	49.73	0.058	7.72	0.744	0.217	1.666	611	0.353	51.40	538 (10 in.)
157–12	0.045	16.33	Neg.	Neg.	0.480	0.165	0.544	365	0.210	16.50	520 (10 in.)
Elgin	0.037	13.50	0.040	5.36	0.179	0.111	0.853	150[c]	0.148	14.49	345 (8 in.)
Clayton	0.027 0	9.85	0.018	2.39	0.117	0.050	0.384	150[c]	0.149	10.43	520 (10 in.)[d]

[a]Baseline flow—dry weather conditions include infiltration.
[b]Inflow is based on a 5-year storm event (1.7 in./hr). Both linear regression and power curve equations were used to fit data to an equation that could be used to predict inflow rates for each area. Reasonable correlations were found to exist between most events and equations. Projected rates were then compared to existing capacity.
[c]Lift stations.
[d]Downstream.

The study area included 21 666 m (71 083 ft) of 200-, 250-, and 300-mm (8-, 10-, and 12-in.) sanitary sewers; 320 manholes; and, based on water consumption records, approximately 1000 customers or buildings. As a result of the low rate of successful access to private property in phase I and a lack of authority to improve accessibility, the District decided to focus phase II on the public sector and to look for sewer capacity solutions to excess I/I problems.

The evaluation survey consisted of the following tasks:

Field investigation

- Flow monitoring at eight locations and rainfall monitoring at two locations from March 14, 1997, to June 3, 1997.
- Manhole and visual pipe inspection of all manholes.
- Smoke testing of all sanitary sewers.
- Internal television inspection of all sewer segments (conducted by the District).

Project management, data analysis, and reporting

- Analysis of system deficiencies.
- Development of a hydraulic model to analyze sewer system operation under existing conditions and with various potential improvements to increase the level of protection from surcharging and backups.
    - Six key sewer line segments were included in the hydraulic model and analyzed under various simulated flow conditions. Most of the reported sewer problems that occurred in May 1996 were along these six sewer segments. Each segment was analyzed for a 10- and 25-year storm event.

- Cost estimates, level of protection, and recommendations.

Findings. The study area is primarily single-family residential with some multifamily residential and commercial developments. The sanitary sewer system studied is inventoried in Table 17.

INFILTRATION CONDITIONS. Typically, the magnitude of infiltration is expressed in milliliters per meter per day per meter (gallons per inch of pipe diameter per mile per day). The average permanent infiltration in the study area is 1.34×10^5 (mL/m·d)/m (1447 gpd/in. diam/mile) and total wet weather infiltration is 3.431×10^5 (mL/m·d)/m (3705 gpd/in. diam/mile). Illinois Environmental Protection Agency design standards specify 1.9×10^4 (mL/m·d)/m (200 gpd/in. diam/mile) as the maximum acceptable infiltration level for new sanitary sewer construction.

INFLOW PROJECTIONS AND ANALYSIS. Table 18 summarizes peak flows for the system for various rainfall events.

Unidentified inflow in Table 19 could primarily be caused by sump pumps, foundation drains, or a combination of both. The 1-year, 30-minute duration-frequency storm, based on rainfall curves for the Rockford area, is one in which a total of 0.95 in. of rain falls in 30 minutes and has a corresponding rainfall intensity of 1.90 in./hr. Total projected inflow based on field inspection activities and flow monitoring for a 1-year, 30-minute storm was 8290 m^3/d (2.19 mgd). Identified public sector inflow was 2060 m^3/d (0.544 mgd), and identified private sector inflow was 167 m^3/d (0.044 mgd). The remaining 6060 m^3/d (1.60 mgd) inflow was unidentified. Table 20 summarizes the defects identified as I/I sources. Although a significant amount of the unidentified inflow is possibly from sump pumps and foundation drains, during smoke testing only five foundation drains were identified as being connected to sanitary sewers.

Conclusions and Recommendations.

PUBLIC SECTOR SOURCE. Removal of identified public inflow sources would remove approximately 25% of total inflow. However, rehabilitation of public sources alone would achieve less than a 5-year level of protection.

PRIVATE SECTOR SOURCES. Locating and removing private sector sources poses some uncertainties. Many building inspection programs have demonstrated that access to at least 90% of buildings is achievable with a well-planned approach. However, it is uncertain whether sump pumps or footing drains or a combination of both are the primary source of private

Table 17 Sewer inventory.

Basin	Size, in.[a]	Footage		No. of manholes
18	4	230	(Private sewer)	
	8	14 293		
	10	466		
Basin total		14 989		70
105	8	10 594		
	10	388		
	12	70		
Basin total		11 052		54
106	8	19 551		
	10	1 370		
Basin total		20 921		93
165	8	9 205		
	10	206		
Basin total		9 411		44
(1/2)107	8	14 710		
Basin total		14 710		59
All	4	230	(Private sewer)	
	8	68 353		
	10	2 430		
	12	70		
System total		71 083		320

[a]in. × 25.4 = mm.

Table 18 Total cumulative flow data summary.

	Basin Number				
Source	107	106	18	165	105
Cumulative base flow					
Average daily base flow	0.046	0.152	0.320	0.053	0.126
Peak hourly base flow[a]	0.078	0.182	0.352	0.074	0.176
Cumulative infiltration					
Peak infiltration	0.060	0.110	0.145	0.055	0.100
Peak inflow					
Peak 1-year, 30-minute storm	0.370	0.725	1.345	0.545	0.845 1
Peak 5-year, 30-minute storm	0.480	0.990	1.820	0.740	1.080
Peak 10-year, 30-minute storm	0.550	1.145	2.115	0.865	1.225
Peak 25-year, 30-minute storm	0.680	1.440	2.680	1.09	1.490
Total cumulative peak flow conditions[b]					
1-year, 30-minute storm	0.508	1.017	1.842	0.674	1.121
5-year, 30-minute storm	0.618	1.282	2.317	0.869	1356
10-year, 30-minute storm	0.688	1.437	2.612	0.994	1.501
25-year, 30-minute storm	0.818	1.732	3.177	1.219	1.766

[a]Average daily base flow multiplied by an average base flow peaking factor.
[b]Peak hourly base flow plus peak infiltration plus peak inflow at the corresponding recurrence interval.
Note: Because of flow split at MN(16)48, it was assumed that 50% of the flow from 107 would flow west through basin 106 and basin 18 and 50% of the flow would flow south through basin 165 and basin 105.

sector I/I. Locating and removing illegally connected sump pumps may not be sufficient to achieve a 10-year level of protection. Approximately 140 sump pumps would account for approximately one-half of the unidentified inflow (i.e., 20% of the buildings in basins 107, 106, and 18). In a recent building inspection program conducted in Rockford, less than 10% of the buildings inspected had illegally connected sump pumps. Footing drains are another unknown factor. The cost of excavating and rerouting footing drains to a storm sewer can be as great as $5500 each. If footing drains are a significant source of unidentified inflow, the removal cost will be significantly greater than the estimated $42 000 for sump pump removal. Locating and removing one-half of the unidentified private sources (sump pumps and footing drains) would not achieve a 10-year level of protection.

Table 19 Comparison of inflow from sources identified during field investigation and total inflow based on flow monitoring projected to a 1-year, 30-minute rainfall.

Basin	Identified inflow, mgd[a]	Inflow, mgd	Unidentified inflow
107	0.107	0.37	0.263
106	0.136	0.54	0.404
18	0.151	0.62	0.469
163	0.086	0.36	0.274
105	0.108	0.3	0.192
Total	0.588 (27%)	2.19	1.602 (73%)

[a]mgd × 3785 = m^3/d.

SEWER CAPACITY INCREASE. Sewer line segment A (which is 250 mm [10 in.] in diameter and lined with a 200-mm [8-in.] P.E. liner) has sufficient capacity to carry large groundwater flows. However, because of the minimum grade and reduced capacity, this line segment acts as a bottleneck to upstream flows, which increases the problem of surcharging caused by I/I. The bypass at manhole (106)48 at the upstream end of line segment A was constructed after the sewer was installed to provide relief to line segment A. However, it has created problems downstream in basins 165 and 105. Increased capacity in line segment A would achieve a level of protection greater than 10 years, almost as great as a 25-year level of protection in basins 107 and 106. A relief sewer parallel to line segment A would not be feasible, and a relief sewer at Cleveland Avenue would unlikely be feasible because of the high risk of potential conflicts with existing sanitary sewers. Removing the 200-mm (8-in.) liner and pipe-bursting the original 250-mm (10-in.) pipe to a 350-mm (14-in.) diameter would be the most feasible method to achieve increased capacity.

Increasing the capacity in line segment A, however, would increase flow rate downstream causing an increase in surcharging and overflows in basin 18. Increased capacity would be necessary from manhole

Table 20 Defects contributing I/I to the system.

Public–private defects	Defect type	Quantity
Public source inflow defects	Pick holes	226
	Ponding in ditch	30
	Broken frame	1
	Frame seal ponding	20
	Frame seals	201
	Corbel leaks	74
	Cracked corbel	17
		Total 595
Public source infiltration defects	Wall leakage	16
	Wall cracked	5
	Brench/trough cracked	12
	Bad pipe seals	22
	Joint infiltration	5
	Pipe cracks	4
	Offset joint	1
		Total 65
Private source inflow defects	Stairwell drain	1
	Foundation drain	5
	Cleanout	18
	Window well drain	1
		Total 25
Private source infiltration defects	Service connection	1

(18)5 to manhole (12)8. Pipe bursting from a 200- to 250-mm diameter was recommended to achieve the added capacity should private I/I rehabilitation prove insufficient. Line segment A is a bottleneck and, regardless of I/I, requires improvement. This is an unusual situation when removing private I/I sources will have only a nominal effect on increasing the level of protection in the study area because of the poor hydraulics of the sewer system. Increasing pipe capacity to achieve a level of protection of 10 years or greater is more costly than a combination of public and private I/I source removal. However, public and private I/I source removal would not achieve the level of protection desired.

Under the existing conditions in the study area, it was recommended that

- The bypass at manhole (106)48 be plugged to achieve a 10-year and greater level of protection in line segment B and basins 165 and 105.
- The capacity of line segment A be increased from 250 to 375 mm (10 to 15 in.) by pipe bursting to achieve a 10-year and greater level of protection in basins 107 and 105.
- The capacity of line segment F be increased from 200 to 250 mm from manhole (18)5 to manhole (12)8 to achieve a 10-year and greater level of protection in basin 18.
- Public sector inflow be removed through manhole rehabilitation.
- Private sector inspections and rehabilitation be completed. (The District has reached agreement with the city regarding private property access authority; so inspections and rehabilitation will proceed.)

APPENDIX A—INFLOW AND INFILTRATION SURVEY QUESTIONNAIRE

Water Environment Federation®
Preserving & Enhancing the Global Water Environment

WATER ENVIRONMENT FEDERATION
SSO COOPERATIVE AGREEMENT WORKGROUP
O&M Monograph Private Sewer Service Committee
Municipality Questionnaire

INSTRUCTIONS

CORRECT MARK	○○●○	• USE A NO. 2 PENCIL, OR A BLUE OR BLACK INK PEN.	• FILL THE OVAL COMPLETELY.
INCORRECT MARKS	⊗ ✓ ▨ ◐	• DO NOT USE RED INK OR A FELT TIP MARKER.	• DO NOT MAKE ANY STRAY MARKS.

PLEASE TRY TO ANSWER THESE QUESTIONS AS THEY RELATE TO YOUR EXPERIENCE WITH PRIVATE SEWER SERVICE CONNECTIONS.
SURVEY FINDINGS WILL BE DISTRIBUTED UPON REQUEST.

A. OWNERSHIP AND ADMINISTRATION

PUBLICLY OWNED:

- ○ Municipal/Township/Regional Government
- ○ Sewer District, Board, or Commission
- ○ Special Regional Authority
- ○ State or Federal Government
- ○ Other: ____________

PRIVATE/INVESTOR OWNED:

- ○ Single Sewerage System
- ○ Subsidiary of Larger Holding Company

ADMINISTRATION: Is your entity also responsible for any of the following? *(Fill in all that apply)*

- ○ Water Distribution
- ○ Wastewater Treatment and Disposal
- ○ Other: ____________
- ○ Stormwater
- ○ Garbage/Sanitation

B. SYSTEM DESCRIPTION

1. Amount of Wastewater Transported by your Entity's Collection System (in million gallons per day):
○ <0.10 ○ 0.10 to 1.0 ○ 1.0 to 10.0 ○ 10.0 to 50.0 ○ 50.0 - 100.0 ○ >100.0
○ Not Applicable

2. Length of Collection System (in miles):

	<10	10 - 50	50 - 500	500 - 1000	>1000	Unknown
Separate Sanitary Sewers	○	○	○	○	○	○
Combined Sewers	○	○	○	○	○	○
Storm Sewers	○	○	○	○	○	○

○ Not Applicable

3. Your entity's Service Area (in square miles):

EXAMPLE
If the amount is 137 square miles, please enter as...***0137***→ (0 / 1 / 3 / 7)

4. Total Number of Employees (dedicated to collection system work, including management and support staff):

EXAMPLE
If the amount is 75 employees, please enter as...***0075***→ (0 / 0 / 7 / 5)

C. SERVICE CONNECTIONS

1. Population Served?

EXAMPLE
If the amount is 1,345,000 served, please enter as...***1345000*** (1 / 3 / 4 / 5 / 0 / 0 / 0)

2. Number of Service Connections:
○ <1,000 ○ 1,000 to 10,000 ○ 10,000 to 100,000 ○ 100,000 to 1,000,000 ○ >1,000,000

3. What type, by percentage, of pipe has been used for private sewer service laterals? (Use best guess)

	Zero	5%	10%	20%	30%	40%	50%	60%	70%	80%	90%	100%
PVC Schedule 40	○	○	○	○	○	○	○	○	○	○	○	○
PVC SDR 35	○	○	○	○	○	○	○	○	○	○	○	○
PVC C900	○	○	○	○	○	○	○	○	○	○	○	○
PVC *other*	○	○	○	○	○	○	○	○	○	○	○	○
Ductile Iron	○	○	○	○	○	○	○	○	○	○	○	○
Cast Iron	○	○	○	○	○	○	○	○	○	○	○	○
Vitrified Clay Pipe	○	○	○	○	○	○	○	○	○	○	○	○
Other	○	○	○	○	○	○	○	○	○	○	○	○

4. What is the minimum private sewer service diameter **currently allowed** in your entity's system?
○ 3-inch ○ 4-inch ○ 5-inch ○ 6-inch ○ 8-inch

FORM 3056 (1/97) 0987654321
DO NOT MARK IN THIS AREA
3335

C. SERVICE CONNECTIONS (cont.)

5. What is the minimum diameter pipe in your entity's system?

○ 2-inch ○ 3-inch ○ 4-inch ○ 6-inch ○ 8-inch

6. Where does your entity require cleanouts? *(Fill in all that apply)*

- ○ At the main
- ○ At bends >45° in lateral
- ○ None required if lateral is less than 100 feet in length
- ○ Other: ______
- ○ Two way at the main
- ○ At the right of way (or easement)
- ○ At bends <45° in lateral
- ○ Every 100 feet along the lateral

7. Does your entity require either of the following for service laterals? *(Fill in all that apply)*

○ Gasketed Pipe ○ Glued Pipe ○ Other: ______

8. Percentage of where private sewer service laterals are located in your entity's system? (best guess)

	Zero	5%	10%	20%	30%	40%	50%	60%	70%	80%	90%	100%	Total 100%
At the back of the property	○	○	○	○	○	○	○	○	○	○	○	○	
At the side of the property	○	○	○	○	○	○	○	○	○	○	○	○	
At the front of the property	○	○	○	○	○	○	○	○	○	○	○	○	

9. What is your entity's definition of a private service lateral (including boundary of entity ownership and limits of responsibility)?

- ○ From wye connection to building
- ○ From right of way (or easement line) to building
- ○ Other: ______

10. Does your entity allow more than one private sewer service connection to one lateral? ○ Yes ○ No

11. Is your entity responsible for maintenance of private service laterals? ○ Yes ○ No

12. Does your entity have jurisdiction to repair/replace service laterals? ○ Yes ○ No

13. If no to question 12, has your entity ever taken responsibility for maintenance or rehabilitation of private service laterals? ○ Yes ○ No

14. Does your entity have a building inspection program which inspects new private service connections? ○ Yes ○ No

15. If yes to question 14, how long has this inspection program been in place?

○ < 2 years ○ 2 to 10 years ○ 10 to 50 years ○ > 50 years

16. Number of inspectors assigned to this task:

EXAMPLE

If the amount is 250 inspectors, please enter as...*250* →

2	0	1	●	3	4	5	6	7	8	9
5	0	1	2	3	4	●	6	7	8	9
0	●	1	2	3	4	5	6	7	8	9

	0	1	2	3	4	5	6	7	8	9
	0	1	2	3	4	5	6	7	8	9
	0	1	2	3	4	5	6	7	8	9

17. Does your entity have a building inspection program which inspects existing private sewer service connections? ○ Yes ○ No

18. Does your entity require inspections of the private sewer service connection at the time of change of title or ownership? ○ Yes ○ No

19. Does your entity require special training or permitting for contractors who perform sewer taps? ○ Yes ○ No

20. Does your entity accurately depict service connections on system maps, drawings or other means such as notebook sketches, construction permits, etc.? ○ Yes ○ No

21. What model of plumbing code is used by your entity?

- ○ Universal Plumbing Code
- ○ Southern States Standard
- ○ Unknown
- ○ Other: ______

D. ADDRESS INFLOW AND INFILTRATION PROBLEMS

1. Which inflow and infiltration (I/I) evaluation methods are utilized by your entity for private properties? *(Fill in all that apply)*

- ○ Night Time (Baseline) Flow Monitoring
- ○ Wet Weather Flow Monitoring
- ○ Pre-and Post-Rehabilitation Flow Monitoring
- ○ Sewer Cleaning Related to I/I Reduction
- ○ Other: ______
- ○ Internal TV Inspection
- ○ Smoke Testing
- ○ Dye Water Flooding
- ○ Building Inspection

2. Which rehabilitation techniques and procedures are used by your entity to address I/I associated with private sewer service laterals? *(Fill in all that apply)*

- ○ Point and Replacement Repairs
- ○ Robotic Point Repairs
- ○ Sewer Lining
- ○ Joint Test and Grout
- ○ Backflow Preventors

3. Is your entity's service area definition based upon...

○ Pump Station Area ○ Gravity Sewershed ○ Pump Station Area and Gravity Sewersheds

4. Is your system defined as a...

- ○ Combined Sewer System
- ○ Dedicated (sanitary only)
- ○ Both (system contains both combined and dedicated sewer sytems)

5. What identifies need and ranking for service area I/I correction/remediation projects? (Assign ranking 1 as top priority through 5 as least priority)

	Rank Order Your Choices				
Gross Flow Rates	1	2	3	4	5
I/I Flow Rates	1	2	3	4	5
Nominal Average Pump Operating Time	1	2	3	4	5
Backups	1	2	3	4	5
Growth/Available Capacity	1	2	3	4	5
Overflows	1	2	3	4	5

D. ADDRESS INFLOW AND INFILTRATION PROBLEMS (cont.)

6. Is inflow/infiltration an important and/or significant consideration when determining which service areas require rehabilitation/remediation during...
   Wet Weather? ○ Yes ○ No   Dry Weather? ○ Yes ○ No

7. When quantifying I/I in a service area, what mini-basin sizes are used? *(Fill in all that apply)*
   ○ Under 10,000 IF   ○ Between 10,000 and 25,000 IF
   ○ Over 25,000 IF   ○ Size is not an issue

8. During an "average" year, what percent of total flow is I/I related?
   ○ 0-20 ○ 20-40 ○ 40-60 ○ 60-80 ○ 80-100

9. What percentage of total inflow/infiltration is related to private service connections?
   ○ <5% ○ 5 to 20% ○ 20 to 50% ○ 50 to 75% ○ >75% ○ unknown

10. Is this percentage based on monitored data? ○ Yes ○ No

11. Please indicate strategy your entity has used to monitor I/I from private service connections, if any:

12. If question 11 was answered, please comment on your perceptions of the cost-effectiveness of your monitoring activities.

13. Do surface water fluctuations have an effect on the inflow and infiltration detected in your system? ○ Yes ○ No

14. Do groundwater elevation fluctuations have an effect on the effect on the infiltration detected in your system? ○ Yes ○ No

15. If yes to question 13 and/or 14, has this been verified by monitoring data? ○ Yes ○ No

16. If yes to question 13 and/or 14, what types of rehabilitation work has your entity performed to address this situation?

17. Do foundation drains tied directly into separate sanitary sewers contribute to wet weather problems experienced by your entity? ○ Yes ○ No

18. How often is groundwater above the crown of most private sewer services?
    ○ Infrequently ○ Most of the time ○ Seasonally

19. Does your entity provide check valves on private sewer service lateral that experience backups? ○ Yes ○ No

20. Please identify any condition unique to your entity's service area that has an effect on the inflow and infiltration detected in your system (e.g., cold climate, geology, hydrogeology, etc.). Explain how your entity addresses this condition in terms of I/I reduction.

21. Does your entity operate any of the following? *(Fill in all that apply)*
    ○ Suction/Vacuum Collection Systems   ○ Other Non-Traditional Collection Methods: ________
    ○ Septage Pumping (Septic Tank Discharge)
    ○ Individual Service Pumping/Transmission

22. How does your entity measure effectiveness of inflow and infiltration reduction efforts?

E. FINANCING AND ENFORCEMENT

1. If your entity rehabilitates private service connections, how is the rehabilitation financed? *(Fill in all that apply)*
   ○ Customer Pays   ○ State Funds   ○ Other Method: ________
   ○ User Fees   ○ Federal Funds
   ○ Local Funds   ○ Reverse/Voluntary Liens

2. If your entity uses any form of liens to finance private sewer service connections, please describe the system.

3. Does your entity use liens as collateral for borrowing? ○ Yes ○ No

4. If the customer pays for private service connection rehabilitation, indicate the arrangements made for payment. *(Fill all that apply)*
   ○ Pay Partial Costs   ○ Payment Plans Available   ○ Other Method: ________
   ○ Loans Available   ○ Pay Up to a Maximum Amount
   ○ Pay Full Costs   ○ Property Liens

E. FINANCING AND ENFORCEMENT (cont.)

5. Has your agency taken any measure to force (or motivate) customers to make repairs to their private sewer service connections? ◯ Yes ◯ No

6. If yes to question 5, please describe the measures taken, and indicate whether it involved surcharges or rate credits and if, in your opinion, the program worked.

____

____

____

____

7. Does your entity have **enforceable** regulations prohibiting downspout/roof drain/area drainage connections into separate sanitary sewers? ◯ Yes ◯ No

8. If yes to question 7, when an illicit connection is found, how does your entity enforce the disconnection requirement?
   - ◯ Penalty
   - ◯ Lien against the property
   - ◯ Fine
   - ◯ Disconnect water service
   - ◯ Nothing
   - ◯ Other:______________________

9. Does your entity permit downspout/roof drain/area drainage connections? ◯ Yes ◯ No

10. Is it legal for one private sewer service lateral to traverse another property? ◯ Yes ◯ No

11. If yes to question 10, are easements required? ◯ Yes ◯ No

F. MANAGEMENT INFORMATION SYSTEM

1. Does your entity use computer technology for its tracking facility information (i.e., computer database)? ◯ Yes ◯ No

2. If yes to question 1, what types of information does your entity manage?
   - ◯ Flow Monitoring Data
   - ◯ Wet Weather Modeling
   - ◯ Preventative Maintenance
   - ◯ Geographical Information System
   - ◯ Other:______________________

3. Does your entity track findings of I/I Evaluations? ◯ Yes ◯ No

4. If yes to question 3, how are findings of I/I Evaluations tracked?

____

5. How does your entity track follow-up work related to I/I reduction recommendations?

____

____

G. PUBLIC RELATIONS

1. Does your entity have a public relations program in place? ◯ Yes ◯ No

2. Are employees of your entity trained in the area of public relations? ◯ Yes ◯ No

3. Does your entity have public education materials related to service connections? ◯ Yes ◯ No

4. Does your entity provide a public awareness program for things such as digging around sewer lines or reporting collection system related problems? ◯ Yes ◯ No

5. What types of procedures are in place for dealing with maintenance work to be done on private property?

____

____

6. What are your entity's procedures for receiving complaints from the public?

____

____

7. What are common complaints your entity receives?

____

____

8. Does your entity have a procedure in place to evaluate common complaints and evaluate the situation? ◯ Yes ◯ No

9. Can your entity provide examples of legislation, public awareness materials, or other information related to private service laterals for use in preparing the O & M Monograph? ◯ Yes ◯ No

10. If yes to question 9, please list materials that may be available:

____

Do you know of any successful I/I Programs that may be of interest when developing the O & M Monograph? ◯ Yes ◯ No

If yes, please list entities and contact names and telephone numbers:

____

____

If you would like a copy of the survey findings please complete the next section.

Name: ______________________ Title: ______________________

Address: ______________________________

Telephone Number: ______________ Fax Number: ______________

E-Mail Address: ______________

Thank you for your time and cooperation!

DO NOT MARK IN THIS AREA

3335

REFERENCES

Carter, W.C.; Nogaj, R.J.; and Hollenbeck, A.J. (1986) Cost Effectiveness and Sewer Rehabilitation. *Public Works,* **117,** Oct, 64.

Clapham, T.W. (1965) The Modern Way To Inspect and Repair Sewers. *Public Works,* **96,** Dec, 92.

Conklin, G.F. (1981) Infiltration/Inflow Removal. *Water Eng. Manage.,* **128,** 4, 34.

Curtis, N.M., and Krutsch, A.H. (1993) Sewer Rehabilitation Using Trenchless Technology on Mainlines and Service Laterals. *Proc. Water Environ. Fed. 66th Annu. Conf. Exposition,* Anaheim, Calif., **6,** 65.

Curtis, N.M., and Krutsch, A.H. (1995) Trying on Flow Reduction for Size. *Water Environ. Technol.,* **7,** 8, 40.

Fedotoff, R.C.; Bellows, P.H.; Frutchey, R.W.; and Thomson, J. (1991) Trenchless Techniques Improve Pipelines. *Water Environ. Technol.,* **3,** 4, 48.

Gray, F.W. (1940) Sewerage in Ancient and Medieval Times. *Sew. Works J.,* **12,** 939.

Hannan, P.; Lovett, R.; and Ralston, T. (1996) Lateral Packers, Grout Close in on Infiltration. *Water Environ. Technol.,* **8,** 3, 45.

Karanik, J.M., and Johnson, P.C. (1976) Conducting a Sewer Inflow/Infiltration Survey To Eliminate 20 MGD. *Public Works,* **107,** Dec, 49.

Kipp, J.S. (1983) Sewer System Rehabilitation by the Numbers. *Public Works,* **114,** Sep, 75.

Lenahan, T., and Herdon, J. (1972) Effective Use of TV Inspection and Sealing. *Public Works,* **103,** Jul, 74.

McCarty, J. (1973) Oakland Maps Program To Combat Sewer Infiltration. *Public Works,* **104,** Feb, 82.

Metcalf and Eddy, Inc. (1930) *Wastewater Engineering: Treatment, Disposal, Reuse.* 2nd Ed., McGraw-Hill, New York.

National Council on Public Works Improvement (1988) Fragile Foundations: A Report on America's Public Works. Final Report to the President and Congress. Washington, D.C.

Paulson, R.I.; Wylie, F.S.; Anderson, D.S.; Miles, F.; and Godfrey, R.T. (1984) Attacking Private Infiltration/ Inflow Sources. *Public Works,* **115,** Feb, 54.

Roberts, G.C. (1979) Expediting Sewer Studies. *Water Sew. Works,* **126,** 2, 52.
Steketee, C.H. (1981) Why EPA's I/I Control Program Fails. *Public Works,* **112,** Jul, 51.
Steketee, C.H., and Beck, R. (1983) Building Sewer Renovation—A Key to Successful I/I Control. *Public Works,* **114,** Jan, 47
U.S. Environmental Protection Agency (1992) *Technical Support Document for Reduction of Pathogens and Vector Attraction in Sewage Sludge.* EPA-822/R-93-021, Office Water, Office Sci. Technol., Washington, D.C.
U.S. Environmental Protection Agency (1996) *Sanitary Sewer Overflows—What Are They and How Do We Reduce Them?* EPA-832/K-96-001, Office Wastewater Manage., Washington, D.C.
Van Natta, W.S. (1963) Fighting Infiltration into Sewers. *Public Works,* **94,** Jul, 158.
Wade, M.G.; Hegwald, D.R.; VanGorden, J.D.; and Fenimore, M.F. (1994) More Than Smoke. *Water Environ. Technol.,* **6,** 3, 61.

SUGGESTED READINGS

Aguiar, L.; Scheller, T.; Cowgill, J.; and Noor, I. (1996) No-Dig Gains Ground. *Civ. Eng.,* **66,** 8, 54.
Curtis, L., and Anderson, G. (1995) Determining Cost Effectiveness of Sanitary Sewer Rehabilitation. *No-Dig Eng.,* **Jun,** 3.
Davis, A. (1998) Research Basis & Recommendations for Private Sector Clean Water Connection Removal Program. Paper presented at Ohio Water Environ. Assoc. Specialty Conf., Columbus, Ohio.
Dillard, C.; Kurz, G.; and Stonecipher, P. (1993) Management of Sewer System Rehabilitation for the Overflow Abatement Program in Nashville, TN. *Proc. Int. Conf. Pipeline Infrastructure II,* Am. Soc. Civ. Eng., New York, 448.
Elliot, J.C.; Hydro, S.K.; Malarich, M.A.; and Weaver, W. (1997) Removing Private Sources of Infiltration and Inflow. *Water Environ. Technol.,* **9,** 8, 55.

Fernandez, R. (1986) Sewer Rehab Using a New Subarea Method. *Water Eng. Manage.,* **Feb,** 28.
Gray, W., and Nogaj, R. (1990) Sanitary Sewer Bypass Reduction Program. *Water Eng. Manage.,* **137,** 5, 36.
Lambert, G., and Hollenbeck, A. (1987) New Approach Achieves Inflow Reduction in Sanitary Sewers. *Public Works,* **118,** 9, 119.
Lateral Sealing System Helps Control I/I (1984). *Public Works,* **115,** Nov, 58.
Macaitis, W., and Kuhl, R. (1994) Local Sewer Rehabilitation—Metro Chicago. *Urban Drainage Rehabilitation Programs and Techniques,* Am. Soc. Civ. Eng., New York, 111.
Maney, S., and Nogaj, R. (1987) Test Program Improves Sewer Rehabilitation. *Water Eng. Manage.,* **134,** 9, 42.
Miles, W.; Gullet, B.; and Zimmer, D. (1996) Managing Old Sewers in the New South. *Water Environ. Technol.,* **8,** 4, 41.
Rowlett, T., and Kelgard, K. (1992) When Sewer Rehab Doesn't Stop Basement Flooding, Water Resources Planning and Management: Saving a Threatened Resource—In Search of Solutions. *Proc. Water Resour. Sessions Water Forum,* Am. Soc. Civ. Eng., New York, 648.
Sewer Rehabilitation Seen as Vital to City's Economic Development (1996). *Public Works,* **127,** 8, 55.
Steketee, C.H., and Heineke, T.L. (1984) The Key to Effective I/I Control. *Public Works,* **115,** Jun, 88.
Stepp, S.G. (1967) Sealing Process Resolves Infiltration Problems. *Public Works,* **98,** Jul, 70.
U.S. Environmental Protection Agency (1975) *Handbook for Sewer System Evaluation and Rehabilitation.* EPA-430/9-75-021, Office Water Program Oper., Cincinnati, Ohio.
U.S. Environmental Protection Agency (1985) *Selected Topics Related to Infiltration and Inflow in Sewer Systems.* EPA-600/2-85-020, Office Res. Dev., Cincinnati, Ohio.
U.S. Environmental Protection Agency (1991) *Sewer System Infrastructure Analysis and Rehabilitation.* EPA-625/6-91/030, Cincinnati, Ohio.

Wells, J. (1994) Service Lateral Reconstruction in the Isle of Man, Infrastructure: New Materials and Methods of Repair. *Proc. Mater. Eng. Conf.,* Am. Soc. Civ. Eng., New York, 257.

Wells, J. (1997) *Service Lateral Rehabilitation Technology Overview, Trenchless Pipeline Projects—Practical Applications.* Am. Soc. Civil. Eng., New York, 143.

Water Environment Federation and American Society of Civil Engineers (1995) *Existing Sewer Evaluation and Rehabilitation.* WEF Manual of Practice No. FD-6, Alexandria, Va.

Wilcoxon, N. (1979) Designing To Relieve Sewer Infiltration Loads. *Public Works,* **110,** Sep, 76.

Wilmut, C., and Sanders, S. (1990) Insights into I/I Investigations. *Water Eng. Manage.,* **137,** 5, 32.

INDEX